T0144539

Digital Audio Forensics Fundamentals

Digital Audio Forensics Fundamentals offers an accessible introduction to both the theory and practical skills behind this emerging field of forensic science. Beginning with an overview of the history of the discipline, the reader is guided through forensic principles and key audio concepts, before being introduced to practical areas such as audio enhancement, audio authentication, and the presentation of reports.

Covering all aspects of audio forensics from the capture to the courtroom, this book is pivotal reading for beginners entering the field, as well as experienced professionals looking to develop their knowledge of the practice.

James Zjalic is a multimedia forensics examiner from Birmingham, UK. He holds a master's degree in Media Forensics from the National Centre of Media Forensics at the University of Colorado, Denver, USA, and a bachelor's degree in Audio Engineering. His casework experience covers the entire scope of the justice system, from criminal through to civil arenas in relation to the most serious of offences and on behalf of various high profile organisations. Research highlights include peer-reviewed publications and international presentations on a range of multimedia forensics topics, including audio authentication and enhancement.

Digital Audio Forensics Fundamentals

From Capture to Courtroom

James Zjalic

Routledge
Taylor & Francis Group

LONDON AND NEW YORK

First published 2021
by Routledge
2 Park Square, Milton Park, Abingdon, Oxon OX14 4RN

and by Routledge
52 Vanderbilt Avenue, New York, NY 10017

Routledge is an imprint of the Taylor & Francis Group, an informa business

British Library Cataloguing-in-Publication Data
A catalogue record for this book is available from the British Library

Library of Congress Cataloging-in-Publication Data
Names: Zjalic, James, author.
Title: Digital audio forensics fundamentals: from capture to courtroom / James Zjalic.
Description: Abingdon, Oxon; New York, NY: Routledge, 2021. |
Series: Audio Engineering Society presents |
Includes bibliographical references and index.
Identifiers: LCCN 2020020024 (print) | LCCN 2020020025 (ebook) |
ISBN 9780367259129 (hardback) | ISBN 9780367259105 (paperback) |
ISBN 9780429292200 (ebook)
Subjects: LCSH: Forensic acoustics. | Digital forensic science. |
Acoustical engineering. | Speech processing systems.
Classification: LCC HV8073.5 .Z63 2021 (print) |
LCC HV8073.5 (ebook) | DDC 363.25/6–dc23
LC record available at https://lccn.loc.gov/2020020024
LC ebook record available at https://lccn.loc.gov/2020020025

ISBN: 978-0-367-25912-9 (hbk)
ISBN: 978-0-367-25910-5 (pbk)
ISBN: 978-0-429-29220-0 (ebk)

Typeset in Sabon
by Newgen Publishing UK

Contents

Acknowledgements

The author would like to extend his thanks to the following people.

My parents, brother, auntie, and grandmother for their continuous and unwavering support throughout my life, my career, and the writing of this book.

Carrie, for encouraging me to pursue my passion for audio and supporting me in whatever I do.

Everybody at the National Center for Media Forensics at the University of Colorado, Denver, USA. I could not have a better academic pedigree in my chosen discipline. A special thanks to Dr Catalin Grigoras and Jeff Smith for their friendship, guidance, and openness in sharing their knowledge and experience.

It should also be noted that some source code made available to me by Dr Grigoras was relied upon for a number of the figures within the authentication chapter.

I must also extend my gratitude to everybody within the audio forensics field who continue to push the discipline forward through their research efforts, and whom without, the field would not exist.

Finally, thank you to Eddy Brixen for his peer review, and the publisher for making this book possible.

Abbreviations

3GPP	3rd Generation Partnership Project
AAC	advanced audio codec
AAU	audio access units
AC	alternating current
AC-3	Acoustic Coder 3
ACPO	Association of Chief Police Officers
ADC	analogue to digital convertor
ADIF	audio data interchange format
ADPCM	adaptive differential pulse code modulation
ADTS	audio data transport stream
AES	Audio Engineering Society
AI	articulation index
AIFF	audio interchange file format
AMR	adaptive multi-rate
ANC	active noise cancellation
ASCII	American Standard Code for Information Interchange
ASF	advanced systems format
b	bit
B	byte
bps	bits per second
BST	British Summer Time
CBR	constant bit rate
CELP	code-excited linear prediction
CRC	cyclic redundancy check
CSD	cross-spectral density
DAC	digital to analogue convertor
dB	decibel
dBFS	decibel full scale
DC	direct current
DCT	Discrete Cosine Transform
DFT	Discrete Fourier Transform
DPCM	differential PCM

DSS	digital speech standard
DSS	differentiated sorted spectrum
DST	Daylight Saving Time
ENF	electrical network frequency
ENFSI	European Network of Forensic Science Institutes
EXIF	exchangeable image file format
FFT	Fast Fourier Transform
FSR	Forensic Science Regulator
GMM	Gaussian mixture model
GMT	Greenwich Mean Time
GPS	Global Positioning Satellite
GSM	Global System for Mobile Communication
HAS	human auditory system
HE	high efficiency
HP	high quality
HVAC	heating, ventilation, and air conditioning
Hz	Hertz
IDFT	Inverse Discrete Fourier Transform
IFF	interchange file format
IID	interaural intensity difference
ILD	interaural level difference
IMRAD	introduction, methods, results, and discussion
IOCE	International Organisation on Computer Evidence
ISO	International Organization for Standardization
ITD	interaural time difference
K	kilo
KBD	Kaiser-Bessel-derived
kbps	kilobits per second
kb/s	kilobytes per second
LC	low complexity
LLR	logarithmic likelihood ratio
LMS	least mean square
LP	long playback
LPC	linear predictive coding
LPCM	linear pulse code modulation
LR	likelihood ratio
LSB	least significant bit
LTAS	long-term average spectrum
LTASS	long-term average sorted spectrum
LTP	long-term prediction
M	mega
M/S	mid-side
MAC	modified, accessed, creation
MAF	minimum audible field

MAP	minimum audible pressure
Mbps	megabits per second
MDCT	modified discrete cosine transform
MFCC	Mel Frequency Cepstrum Coefficient
MIPS	millions of instructions per second
MOS	mean opinion scores
MP3	MPEG Version 1 Layer 3
MPEG	Motion Picture Experts Group
MSB	most significant bit
MSE	mean squared error
PCM	pulse code modulation
PESQ	perceptual evaluation of speech quality
PNS	perceptual noise substitution
POTS	plain old telephone system
PSD	power spectral density
PSTN	Public Switched Telephone Network
PTS	permanent threshold shift
RIFF	resource interchange file format
RIR	room impulse response
RMS	root mean square
RTF	room transfer function
SBR	spectral band replication
SI	international system of units
SNR	signal to noise ratio
SOP	standard operating procedure
SP	standard playback
SPL	sound pressure level
SSR	scalable sampling rate
STI	Speech Transmission Index
SWGDE	Scientific Working Group for Digital Evidence
TDOA	time difference of arrival
TNS	temporal noise shaping
TOA	time of arrival
TTS	temporary threshold shift
UCTE	Union for the Coordination of Transmissions of Electricity
UPS	Uninterruptible Power Supply
UTC	Coordinated Universal Time
V	volt
VAD	voice activity detection
VBR	variable bit rate
VOIP	Voice Over Internet Protocol
W	watt
WAV	waveform audio file format
WAVE	Windows Audio File Format
WMA	Windows Media Audio

Chapter 1

Introduction

In the modern world, the word 'Forensics' is defined as the application of science to the justice system, but the actual inception of this word harks back to 44 BC and the post-mortem examination of the infamous Roman leader, Julius Caesar. Upon performing his investigation, the physician, Antistius, arrived at the conclusion that Caesar had been stabbed a total of 23 times, and that it was the second piercing of the chest that inflicted the fatal blow. Once his analysis was completed, he was asked to present his findings 'before the forum,' an archaic version of the court system used across the world today. As such, the word itself is derived from the Latin term 'forensis' or 'of the forum' when translated into English (Wilkinson, 2015). Since the coining of its name, many sub-disciplines have since evolved, with the most recent addition being digital forensics. As it relates to the pertinent topic for this book, digital audio forensics is, therefore, the application of the science of digital audio to legal matters.

In recent years there has been an increase in the exposure of this very niche field of forensics, through real-life cases covered by the media, true crime podcasts and shows which grace our cinema and television screens. The reasons for this newfound publicity are twofold and can be surmised to be due to the proliferation of media outlets combined with the increased use of audio forensics within the courtroom. The latter is primarily a consequence of the now ubiquitous nature of digital devices consuming every waking hour of our lives, the majority of which are capable of capturing sound.

With this in mind, it is essential that accurate and reliable information exists on the subject, not only to allow the next generation of audio forensic practitioners to have the prerequisite information to continue to serve the justice system but also to provide those who are looking for further education in this discipline to have material presented in a non-biased, objective, scientific but accessible format. Although the vast majority of the information contained within this book is available from other sources, many require subscriptions to scientific journals, a large library, extensive in the field training or a rare educational programme.

Scope

This book is written with the intention of being both a primer for those interested in the topic, and as a reference guide for those already active within the field of audio forensics. As such, attempts have been made to ensure it contains the necessary information for a reader to understand all themes covered through the gradual cumulation of knowledge in a progressive manner.

References are provided at the end of each chapter, should further knowledge relating to a specific concept be required. Those with an understanding of general audio concepts which relate to audio forensics may wish to advance to chapters which are of particular interest to them, but it is recommended that the book be read from cover to cover so concepts for which *a priori* knowledge is required can be reviewed to aid in understanding the preceding chapters. The presentation of information is designed to allow the book to serve as a reference for those within the field, and attempts have been made to minimise jargon. Where technical concepts are used, they are dissected to aid understanding.

Although the term audio forensics can be applied to both digital (data whose foundation is a binary representation of ones and zeros) and analogue (everything else which is captured in a non-digital format, for example, vinyl records and magnetic tape), there are as many differences between the two formats as there are similarities. These differences will be touched upon in Chapter 2 to enable the reader to gain perspective from the history of audio forensics and learn how the transition from analogue to digital audio came about, but to encounter analogue audio when working in forensics is now extremely rare as the majority of modern systems are digital. The only cases pertaining to analogue audio will, therefore, be appeals and cold cases. In light of such, it is solely the digital domain on which this book is focused. It appears that the digital element is here to stay, thanks to the lower costs, reduced physical size, increased storage capacities, and the ability to create perfect copies of data among other advantages.

There are also several areas in digital audio forensics which cross into other disciplines, including voice comparisons and analysis. This field lies somewhere between phonetics and statistics, and digital audio is the medium in which the pertinent data is stored. This area will be covered briefly for the sake of completeness, but for detailed information concerning such a topic, other sources by more suitably qualified experts should be sought. Acoustic analysis of signals is another area which is extremely broad and requires a deep understanding of both physics and various mathematical concepts. As the book is to be considered an introduction, an overview will be given with regards to the area, but, as with voice comparisons and analysis, further sources of information on the subject should be consulted.

When people are first introduced to audio forensics, comments are generally made which concern recordings captured via telephones or mobile recording devices, but this is only half the story. Video recordings are essentially a stream of images synchronised with an audio track, and although audio recordings in and of themselves are ubiquitous, the video format is probably the most popular multimedia format in the world as it provides both a visual and audible capture of events. There are, therefore, a phenomenal number of recordings captured every second of every day, and so it is somewhat inevitable that some will end up capturing evidence of an event which can be then used in a legal capacity.

Although scientific principles are the same throughout the world, the judiciary systems and protocols for presenting evidence are not (Ambos, 2003; Block et al., 2000; Froeb and Kobayashi, 2001). The non-technical information provided will, therefore, be agnostic to legal systems, regions, and countries, and will only include areas which pertain to the majority of the population.

Finally, although specific tools will be referenced where used to create demonstrative imagery in the form of figures, the book is software agnostic for two reasons. First, there are no analyses or processes within audio forensics which are only available by using a single tool or piece of software. In reality, there are numerous tools which can perform the same analysis and render the same results. It would also be a disservice to provide a 'how-to' guide on a piece of software without explaining the science behind the processes. Far more important is the understanding of concepts so a reader can use any tool they find to be suitable (or create their own tools), perform their own research, and present findings and answer questions in a scientific and logical manner. Understanding why a process is applied, and the impact of the said process on the audio signal is what separates forensic scientists from those who know which buttons to press but have no idea why they are pushing them, what is happening when they do, and what the output really means. A lack of understanding can lead not only to poor analysis through the loss of essential data (in the case of enhancement) but also to misinterpretation (in the case of audio authentication and analysis) and potentially wrongful releases or convictions.

Foundations

Just as a tree is only as strong as its roots, foundations must first be formed to provide a stable structure to all that follows. In order to achieve this, the audio forensics discipline must, therefore, be traced all the way back to the phenomenon to which every living (and non-living) thing is continuously exposed: Radiation (Lamb, 1910; Strutt and Rayleigh, 1877).

Radiation exists all around us in the form of frequencies which can be perceived directly by humans, such as light, heat, and sound, and those for which we require a medium to transpose it into a form which we can identify. For example, although dolphin chatter exists in a spectral region beyond the limits of our auditory system, these sounds can be transposed into those which we can discern. X-rays must be converted to a visual representation for analysis. Microwaves (the kitchen appliance) take their name from microwaves, which we cannot directly sense, but can feel as heat through the medium of food. When all the other forms of radiation are considered, we can hear only an extremely thin slice of the entire spectrum (Figure 1.1).

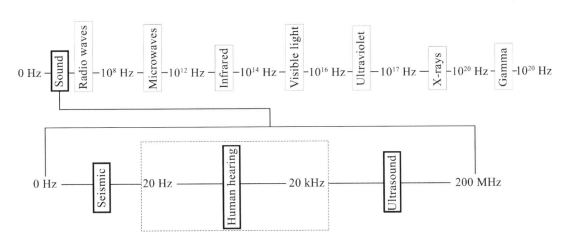

Figure 1.1 The frequency bandwidth perceptible to humans within the electromagnetic spectrum

Figure 1.2 The conditions required for audio forensics

For thousands of years, sound existed as a transient event, ceasing once the vibrations perceived by the human auditory system had passed. Although the events heard in this manner were, and still are, used within the justice system as evidence in the form of utterances heard by witnesses or conversations heard between two parties, it is highly subjective. Perceived sounds are not only extremely susceptible to human bias (both conscious and subconscious), but also to other factors such as Post-Traumatic Stress Disorder. In addition to this, the sounds heard by a witness cannot be reproduced. This does not, therefore, lend itself to any form of analysis, and, as such, is of limited use and reliability when compared to other types of evidence. Once advances in recording had reached the point that sound could be captured and stored, it was in a form in which the aforementioned factors were no longer an issue and could be replayed and analysed in an objective and scientific manner.

For a sound to become potentially evidential, there are several conditions which must coincide. The first is the occurrence of an event (typically a crime or something relating to a crime). The second is the propagation of an acoustic wave pertinent to the event (for example, a verbal confession or a gunshot). Providing somebody is present, this has some potential as evidence in the form of an individual perceiving the sound and recalling this to a trier of fact, but as discussed above, this would be subjective, due to flaws in the human memory and potential biases. In light of this, there have been numerous studies which have shown the effects of stressful incidents on the mind, even when recollecting directly after the event (Clifford and Bull, 1978; Cook and Wilding, 1997, 2001; Mullennix et al., 2011; Pankkonen et al., 2017; Paunovic et al., 2002; Wells, 1995). When one considers that the witness may not be providing evidence within a courtroom until after an extended period following the incident, the memory is likely to have degraded even further, and subsequently, the reliability of the evidence is further reduced. Biases may also exist, either consciously or sub-consciously. If the witness is an acquaintance of the subject involved, biases may result in them giving an account which supports their friend's position, or does not provide a full account of what really occurred. Subconsciously, they may also have inherent biases, all of which may distort the evidence. The biases essentially mean that this type of evidence is not too reliable. This is where the meeting of the third condition, that of the capture of propagated sound as an audio recording, becomes critical. If this condition is met, a snapshot of the acoustic events which occurred has been created (Figure 1.2). A witness statement can then potentially be replaced, verified, or challenged by the ironic 'silent witness' of a digital audio recording.

The first two conditions can be reasoned to have a relatively high possibility of occurring based on Locard's principle of exchange. Dr Edmund Locard (1877–1966) was a French scientist, who, in 1910, built a laboratory in the French city of Lyon, and is considered a pioneer of forensic science. His theory forms one of the two pillars of forensic science, expressed by (Miller, 2014) as:

The truth is that none can act with the intensity induced by criminal activities without leaving multiple traces of his path. The clues I want to speak of here are of two kinds: Sometimes the perpetrator leaves traces at a scene by their actions, sometimes, alternatively, he/she picked up on their clothes of their body traces of their location or presence.

Once this theory took root and was popularised in 1930s North America, it became simplified as 'Every contact leaves a trace' (Roux et al., 2012). P.L. Kirk, a professor at the University of California Berkeley Criminology School later expanded on this theory, writing:

Wherever he steps, whatever he touches, whatever he leaves, even unconsciously, will serve as a silent witness against him. Not only his fingerprints or his footprints but his hair, the fibers from his clothes, the glass he breaks, the tool mark he leaves, the paint he scratches, the blood or semen he deposits or collects. All of these and more, bear mute witness against him. This is evidence that does not forget. It is not confused by the excitement of the moment. It is not absent because human witnesses are. It is factual evidence. Physical evidence cannot be wrong, it cannot perjure itself, it cannot be wholly absent. Only human failure to find it, study and understand it, can diminish its value.

(Kirk, 1963)

Kirk also coined the term 'Individuality Principle,' which can be considered to be the second pillar of forensic science. The theory is based on the assumption that 'every object in the universe is unique' and that forensic science should focus on the source of two items (the questioned and the known). Further to this, one of his PhD students (Kwan, 1977) proposed that the agreement between properties measured from a trace (e.g., a fingerprint found at the crime scene) against a source (e.g., a fingerprint taken from a suspect) did not necessarily mean they were left by the source. The Bayesian approach proposed by the author for use in identification processes is still used across a number of forensic disciplines for the reporting of the results of an analysis to this day.

In the latter half of the twentieth century, the possibility of audio pertaining to a crime being captured became increasingly likely as law enforcement agencies focused on the recording of sound using wiretaps and bugs. In the present age, this possibility has further expanded due to the proliferation of audio recording devices, most notably in the form of smartphones, portable digital recorders (including those used covertly), and call centre recording systems.

As the capture of audio became more likely, developments in the field of audio analysis were also picking up steam. The first use of audio analysis was born out of necessity during the First World War when the US-based Bells Labs created a system to identify submarines. Once the war was over, a company named Kay Electric made the product commercially available, and in 1951 trademarked and marketed it as 'The Sona-graph,' producing graphs known as 'Sonograms' (Vale, 2019). It is this conversion of audio from the time domain (in which a recording is simply represented as changes in amplitude over time) to the frequency domain which gave birth to audio forensics as we now know it. Conversion to a frequency-domain representation allowed recordings to be visualised as individual frequencies, opening up a whole new world of possibilities for processing and analysis. Although the spectrogram is still used to this day, it is the analyses and processes born out of the frequency representation which have paved the way for researchers to propose more innovative ways of maximising the potential of audio recordings for use in forensics. The succeeding emergence of digital computing and digital signal processing exponentially increased audio analysis possibilities due to their ease of use, the speed at which processes can be performed, and the methods of visualisation available.

It is the culmination of these events that have led us to the point at which we find ourselves in the present day, one in which audio forensics has become both commonplace and essential to judiciary systems around the world.

Sub-disciplines overview

Although a niche field in and of itself, digital audio forensics also consists of a number of sub-sections, some of which overlap with other disciplines. Although each will be discussed in detail in its own chapter, the following can be considered a brief overview of the various sectors.

Research

Without the research performed by companies such as Bell Labs during the Second World War, through to that published in scientific journals and presented at forensics conferences annually to this day, the field of audio forensics would either not exist, or would not afford us the variety and reliability of analyses we have at our disposal. Research is the cornerstone of forensics, and this is a point which cannot be emphasised enough. It is essential for all of the sub-sets which follow and in giving examiners the means to stay one step ahead of nefarious parties so their work can be of benefit to the justice system. It is rare for private forensic laboratories to perform research as casework is their focus, and as commercial entities, their primary objective is profit. It can, therefore, be challenging for them to find the funds or the time. Much of the burden of research is consequently shouldered by university departments, governmental organisations, and forensic software development companies around the world. Once the methodologies and results have been subjected to peer review and subsequent publication, they are then adopted by forensic practitioners in real-world cases. The use of non-peer-reviewed techniques is not only frowned upon but is extremely dangerous due to the unknown error rates and potential unreliability which could result in wrongful conviction or release.

Capture

The capture of audio can be a field in and of itself, especially within law enforcement and government security services. This area covers the use of various devices to capture sound in the hope that information relating to a crime will be captured. Many of the methods will be familiar due to their inclusion in movies and TV shows, for example, an informant 'wearing a wire' and attempting to coerce a confession from a criminal, or the use of 'wire-tapping' to record a conversation between two criminals talking over the phone. Today wire-taps are most commonly associated with older movies involving three-letter organisations in the US. The name comes from the use of a hardware device to tap into the physical wiring which connects a phone to the outside world, something which came with its own challenges of obtaining physical access while being undetected. Since the emergence of the digital era, the use of physically connected phones has decreased, and the majority of telephony communication is performed digitally. This lends itself to the capture of telephone recordings without the need for a hard-wired physical connection. Wire-tapping and many similar methods were first developed during the days of analogue audio (hence the 'wire' prefix), but have survived the leap to the digital age through the use of updated techniques.

In days gone by, a covert recording captured by a subject 'wearing a wire' would involve concealing a device attached to a microphone by a wire on the person. In the present day, this is much easier due to the possibility of transmission over a wireless network and the reduction in size of devices eliminating bulky hardware. Although essential for real-time monitoring of events, wireless communication does come with its own problems. Tools are available to detect the transmission of radio waves from the capture device, the waves may be blocked in some environments, and the use of wireless communication is not as safe as a hard-wired capture due to the possibility of signal dropouts, electrical interference, and interception of the signal. The safest option is, therefore, to store the captured audio to a device on the person. This has been made easier through the availability of smaller components which have allowed recording devices of various guises to enter the market, such as those hidden within pens and keyrings through to those smaller in dimensions than the smallest of coins. This kind of covert recording does not necessarily need the microphone to be worn and could be placed in a static location, for example, a pen could be placed in a pen holder among genuine pens, or a recorder could be hidden under a table. Some of the more interesting devices include dummy ceiling tiles where original tiles are replaced by ones containing microphone arrays and installed above a table where a meeting is scheduled to take place. Another example is a laptop cover used with the laptop open and the array of microphones directed towards the unsuspecting victim.

One further recording method worth mentioning is 'dial-in probes.' These are devices which are placed within an environment and contain both a microphone and a sim-card (basically a no-frills mobile phone minus the speaker). A user can dial the number pertaining to the device, activating the microphone and thus allowing recording to take place.

For all forms of covert recording, there is a level of technical skill required to ensure a device with sufficient specifications for the task at hand is used, the positioning of the microphone is optimised, and the potential for noises being captured which may mask the desired audio is minimised.

Authentication

Maintaining the integrity of evidence is one the most important areas of forensic science as it serves to ensure the most original manifestation of the evidence is that which is considered by the examiner, and therefore the court. This transcends all fields of forensic science, whether it is ensuring there is no cross-contamination during DNA analysis, or no additional markings made to a bullet submitted for ballistic examination. In the case of audio recordings, its purpose is to ensure the recording has not been manipulated or converted to a new format which may reduce the quality or integrity, potentially resulting in misinterpretation or a loss of information. If a recording which had been edited was relied upon by the court or by an examiner during an examination, all results and interpretations thereafter would be unreliable. In the most extreme case, this would be the creation of a recording through the use of voice synthesis software to frame somebody by having them appear to say things which they did not actually say. More innocent reasons may pertain to somebody trimming a recording to reduce the file size so it can be transmitted as an email attachment, removing potentially relevant information either side of the cropped area in the process. It is essential to the criminal justice system that all findings are based on the most original recording to prevent any form of misinterpretation by the triers of fact, regardless of its importance to a case. For this reason, authentication examinations consisting of various analyses are performed to provide an opinion as to the provenance of a recording and the possibility of editing having taken place.

Enhancement

Audio enhancement is defined by SWGDE (the Scientific Working Group for Digital Evidence) in their *Digital and Multimedia Evidence Glossary* (SWGDE, 2016) as: 'Processing of recordings for the purpose of increased intelligibility, attenuation of noise, improvement of understanding the recorded material and/or improvement of quality of ease of hearing.'

Forensic audio recordings are, more often than not, captured in noisy environments using low-quality equipment. This can result in the presence of various artefacts, some of which are superfluous to the desired content. Imagine an individual speaking in a room where the sound emitted from their mouth bounces off the highly reflective material of which the walls, floor, and ceiling are constructed causing indirect reflections to smear words or a gunshot captured in a drive-by shooting. Both contain information which is detrimental to the desired signal. In the first instance, reverberation, and in the second, the noise of the engine and tyres against the road. Using various methods, this non-essential data can be reduced, leaving only that which is required before then further improving the desired signal.

Acoustic analysis

Although the majority of forensic audio work requires some form of analysis, acoustic analysis pertains to the physics of sound and makes use of the content of the recording relating to the capture environment by using acoustic theory to perform analyses and draw conclusions. This may be the location the recording was captured, the identification of gunshots, or the number of times an individual was struck with a weapon. The acoustic representation of a sound is influenced by two factors: (1) the environment in which it was captured (and thus reflections, microphone positioning and shielding); and (2) the recording device itself. An understanding of both is required to ensure there are no misinterpretations relating to the cause of specific findings during analysis.

Voice analysis

When people think of audio forensics, it is probably the analysis of speech which first comes to mind, namely, voice comparison and speaker recognition. Voice comparison examinations relate to the comparison of voices to aid in providing an opinion concerning the level of similarity between them. This requires not only an understanding of digital audio (so the effects of capture can be understood) but even more so, phonetics and statistics. Speaker recognition is used for biometric voice systems to verify the identity of a speaker, and is therefore not applicable to audio forensics, although both employ many of the same techniques.

Speech recognition, similar to speaker recognition, is not generally used within digital audio forensics and is also more commonplace in biometrics, for example, in mobile phones and smart home devices. It is used to determine *what* was said by modelling an individual's voice, rather than *who* said it. As forensic audio recordings are often captured in noisy environments in less than ideal conditions, speech recognition has little use in this field. Speech analysis is used to determine what was said, for example, did the business owner say 'I'm going to *bill* him for that' or 'I'm going to *kill* him for that' about a client who ended up murdered? In the majority of cases, the audio will be of an intelligibility which can be transcribed by an individual with an understanding of the audio capture process (to differentiate between distortion and speech) using high-grade equipment in ideal listening conditions. Analysis of disputed utterances is required when (1) the statement is considered essential or (2) the audio is of a quality which is not easily understood, or, in many cases, both. In this case, as with speaker comparison, a phonetician with an understanding of digital audio capture is required.

Summary

For audio forensics to be possible, three simultaneous occurrences are required, namely, an event (for example, a crime), a sound to be propagated in relation to the crime (for example, a gunshot), and for the sound to be captured by a recording device (for example, a mobile phone). Although audio forensics is a niche field, once a recording reaches an audio forensics laboratory, a broad range of analyses are available, including enhancement, authentication, and the comparison of voices. As every case is different, the application of each analysis is dependent on the audio recording itself and the requirements of the instructing party, but when performed correctly, the results and subsequent conclusions can provide both compelling and extremely reliable evidence.

References

Ambos, K., 2003. International criminal procedure: 'adversarial', 'inquisitorial' or mixed? International Criminal Law Review 3, 1–37.

Block, M., Parker, J., Vyborna, O., and Dusek, L., 2000. An experimental comparison of adversarial versus inquisitorial procedural regimes. American Law and Economics Review 2, 170–194.

Clifford, B.R. and Bull, R., 1978. The psychology of person identification. Routledge, London.

Cook, S. and Wilding, J., 1997. Earwitness testimony: Never mind the variety, hear the length. Applied Cognitive Psychology 11, 95–185.

Cook, S. and Wilding, J., 2001. Earwitness testimony: Effects of exposure and attention on the face over-shadowing effect. British Journal of Psychology 92, 617–629.

Froeb, L.M. and Kobayashi, B.H., 2001. Evidence production in adversarial vs. inquisitorial regimes. Economics Letters 70, 267–272.

Kirk, P.L., 1963. The ontogeny of criminalistics. The Journal of Criminal Law & Criminology 54, 235–238.

Kwan, Q.J., 1977. Inference of identity of source. University of California Press, Berkeley, CA.

Lamb, H., 1910. The dynamical theory of sound. Edward Arnold, London.

Miller, M.T., 2014. Locard's exchange principle, in Crime scene investigation laboratory manual. Academic Press, Oxford, pp. 15–20.

Mullennix, J.W., Ross, A., Smith, C., Kuykendall, K., Conard, J., and Barb, S., 2011. Typicality effects on memory for voice: Implications for earwitness testimony. Applied Cognitive Psychology 25, 29–34.

Pankkonen, O., Kiiskinen, K., Kaakinen, J.K., and Santtila, P., 2017. Understanding of and memory for a complex auditory event: An experimental case study to resolve an evidentiary issue in a trial. The Journal of Forensic Psychiatry & Psychology 28, 70–90.

Paunovic, N., Lundh, L.-G., and Öst, L.-G., 2002. Attentional and memory bias for emotional information in crime victims with acute posttraumatic stress disorder (PTSD). Journal of Anxiety Disorders 16, 675–692.

Roux, C., Crispino, F., and Ribaux, O., 2012. From forensics to forensic science. Current Issues in Criminal Justice 24, 7–24.

Strutt, J.W. and Rayleigh, B., 1877. Theory of sound. Macmillan and Co, London.

SWGDE, 2016. Digital and multimedia evidence glossary. Version 3.0. Available at: www.leva.org/wp-content/uploads/2019/10/SWGDE-Glossary.pdf

Vale, A., 2019. Sonograph. A cartooned spectral model for music composition. Paper presented at the Sound, Music and Computing Conference, Malaga, Spain.

Wells, G.L., 1995. Scientific study of witness memory: Implications for public and legal policy. Psychology, Public Policy, and Law 1, 726–731.

Wilkinson, C., 2015. A review of forensic art. RRFMS 5, 17–24.

Chapter 2

The history of audio forensics

Technological developments

The analogue years

Present-day digital audio forensics is a culmination of technological advances in the capture and analysis of sound propelled by a few significant events in which it was required as evidence in legal proceedings.

Although Thomas Edison is widely regarded as a pioneer in the field of audio, the first capture of sound is documented as the 10-second recording of a folk singer on the 9th of April 1860, in France. The device was named the Phonoautograph and consisted of a barrel-shaped horn attached to a stylus. The movement of the horn, and thus, the stylus, etched sound waves onto sheets of blackened paper which had previously been exposed to oil lamp smoke. Designed by Parisian Edouard-Léon Scott de Martinville, it was his intention to capture a visual representation of human speech that could be deciphered at a later date (Brock-Nannestad and Fontaine, 2008), and for around 150 years the sounds remained trapped on the blackened paper. But with new technology comes new solutions, and in 2008 scientists at the Lawrence Berkeley National Laboratory in California, USA, used optical imaging on the high-resolution scans of the captures, and the first recorded sound was finally replayed (Rosen, 2008).

During the late nineteenth century, research interest began to focus on improving communications by creating a system which could transmit speech from one region of the USA to another. This was partially achieved on the 14th of February 1876, when Alexander Graham Bell, in his laboratory in Boston, MA, used a diaphragm attached to an electromagnet to act as a transducer and convert acoustic waveforms hitting the diaphragm into electrical variations. The speech was extremely faint, but audible none the less (Clark, 1993). He called his new device the 'telephone' after the word 'telephony' (which was already an established term by this time). It should be noted that although Bell is considered the inventor of the telephone, an individual named Philipp Reis from Germany had presented a lecture on his 'telephone' in 1861, which was shown to work over a distance of 200 metres. Unfortunately for Reis, nobody showed any significant interest at the time (Voelker and Fischer, 2002).

A year later and upon hearing of Bell's device, Thomas Edison took this idea one step further with the intention of transmitting messages using this new technique. Using a stylus attached to the telephone microphone, he was able to create grooves representative of the movement of the sound waves on paraffin-coated paper by using an electromagnetic as a transducer (Figure 2.1).

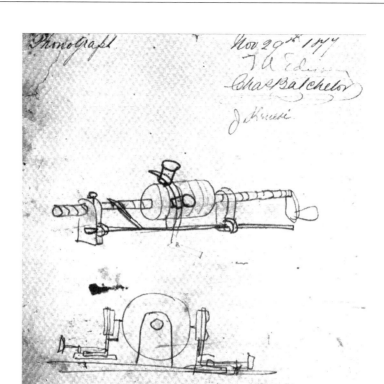

Figure 2.1 Thomas Edison's first sketch of the Phonogram, dated November 29th 1877

He discovered that reversing this process by running the stylus over the grooves moved the connected electromagnet, and thus the diaphragm. The diaphragm movement then moved the surrounding particles of air, reproducing the previously captured sound waves of the nursery rhyme 'Mary had a little lamb.' He named his device the 'Phonograph' (Bouchard, 2012; Wile, 1982). The concept of reversing the capture process for playback remains the foundation of all audio systems to this day.

While Edison moved onto other projects, Bell picked up on the work of his rival and swapped the stylus for a chisel-shaped cutter to leave deeper, and thus cleaner grooves. This resulted in a more intelligible sound with lower background noise. He named his device the 'Graphophone' in his 1886 patent (Newville, 1959). Although the two men were clearly competitive and spurred on by one another's inventions, Bell decided to approach Edison with the proposition of the two working collaboratively. This was rejected by Edison, who was not in favour of somebody else improving on his inventions. It also reignited his competitive flame and propelled him back into his work on the phonogram. This time he decided to use electric motor-driven disks instead of the hand-cranked rotating cylinder he had used previously for movement of the paraffin-coated material and he named his new 1898 device the 'Gramophone' (Collins and Gitelman, 2002).

While some were focusing on the entire process of capture and playback, others were researching specific elements. Valdemar Poulsen described the separation of the carrier from the memory coating in a flat information medium in a 1900 patent, an observation which went on to form the basis of magnetic tape (Poulsen, 1900). He also recognised that the signal captured on a tape does not correspond to the recording current but is distorted, and in 1907 he proposed the addition of a DC current to the signal, a technique known as 'biasing.' The method pre-magnetises the magnetic medium to reproduce the magnetic flux from the recording head more precisely, thus reducing distortions (Poulsen and Pederson, 1907). E.C. Wente, of Bell Telephone Labs, published a research article (Wente, 1917) describing a uniforming sensitive instrument for the 'absolute measurement of sound intensity,' named 'the condenser microphone.' Although initially designed for long-line telephone transmission during the USA's development of a transcontinental telephone service, its remarkable acoustic performance lent it to the recording process (Mayfield and Harrison, 1926).

Although these early inventions were revolutionary and indeed transformed the world, their key limitation was the modest magnitude at which sound could be replayed. This was due to the transducer components of the early devices moving in direct physical relation to the sound waves, resulting in a small amount of movement and a subsequently faint sound. It wasn't until Western Electrical performed research into the amplifier, a device designed to increase the strength of electrical currents, that the magnitudes of the playback signal increased and a much more extensive range of frequencies could be produced. This was made possible as the higher electrical current allowed the cutter attached to the electromagnet to make grooves in a plastic record rather than on foil or paraffin-soaked paper. These grooves were more pronounced, and as the process was reversed on playback, the movement of the speaker's diaphragm could also be increased, displacing surrounding air particles to a more considerable degree.

In 1924, Western Electrical took the amplifier on tour around the USA to various recording studios, resulting in the use of the amplifier in a device called the 'Orthophonic Victrola' (Figure 2.2) (*The New York Times*, 1925).

A year later, the moving coil transducer was developed, consisting of a coil of thin wire attached to a diaphragm dangled within a magnetic field. This technique relied upon movement of the diaphragm (and thus the coil) to change the electric current by corresponding amounts. It was so far ahead of the competition that, almost overnight, all other microphones disappeared, essentially made redundant by this new invention (Maraniss, 1937).

As the recording industry embraced the new technology, the film industry began to take notice and started building on the developments made by Western Electrical. Not only did they desire higher levels of capture, but also higher levels of playback in order to fill cinemas with sound. This research resulted in the three-way speaker in 1931, for which splitting frequencies into three distinct bandwidths was at the heart of its success. The bands consisted of the woofer (containing low-end frequencies), the mid-range driver (mid-range frequencies), and the tweeter (high-end frequencies). This splitting concept is still used for speaker design and construction to this day (Millard, 2005).

By this point, research into the potential of magnetic tapes for the capture of sound was well underway. In 1928, one of the first machines that could be used with magnetic tape was developed by Fritz Pfleumer, in which paper coated with steel shavings served as the storage method. This brought in the potential for looping, splicing, and erasing of audio content for the first time

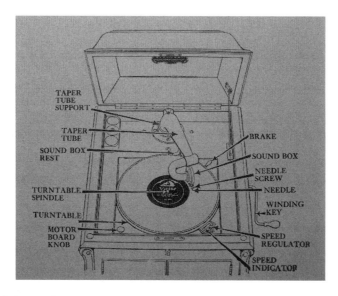

Figure 2.2 Victrola XI diagram from the original instruction manual
Source: Many thanks to the Antique Phonograph Society for supplying this image.

in the history of recording. Upon realising he could not take his invention any further without the help of a larger organisation, Pfleumer approached AEG, who purchased the patent for his invention from him in 1932. Although AEG began developing a tape machine, they were primarily an electronics company with little knowledge of how to make the tape, so they outsourced the manufacturing of this to the chemical company BASF, who at the time were known as IG Farben (Thiele, 1989). The collaboration between the two companies paid off, and, in 1935, AEG presented the Magnetophon K1. This was followed by the first production model called the Magnetophon K2 in 1936 (Krause, 2002). Although the developments made in magnetic tape were seen as a success, there was still an audible amount of distortion upon playback. The DC bias solution proposed around the turn of the century by Poulsen was further developed in 1938 by Dr Walter Weber who achieved 3 dB of dynamic improvement by sending an AC current (and thus creating an AC bias) to the record head of the machine at the same time as the audio signal. He then continued to refine the technique, and in collaboration with Dr von Braunmuhl, the duo patented 'A Process for Magnetic Tape Recording' in 1940. The increase in quality that it provided led to the purchase of the patent by AEG, who then implemented the technology into all of their Magnetophon machines from 1941. Once film and music recording studios began using the devices, their place in history was solidified (Thiele, 1990).

By the 1940s, this technology was finding some use in the military due to contracts issued by the Naval Research & Development Department. While the Allies used magnetic tapes during the Second World War, Hitler's demand for audio to be replayed for propaganda purposes meant Germany continued its research into the field. This led the German company AEG to develop a tape using paper or acetate covered with a fine layer of brown iron oxide, a method which was to go on to become the future of magnetic recording (Engel and Hammar, 2006). It wasn't until the final months of the war that this invention was discovered, and once the war was over, the US took two Magnetophon machines back to a laboratory and performed reverse engineering to recreate their own versions (Audio Engineering Society, 2014).

As is often the case, the urgency initiated by wartime conditions caused many developments to be made during those years, and the commercial manufacturing and marketing of products such as magnetic tape, vinyl records, and stereo recordings did not happen until after the war. Some nine years after the war ended, major record companies finally started selling their material on a pre-recorded tape.

Although revolutionary at the time, magnetic recording was not without its problems. Reel-to-reel tape was large, and the need to thread the tape through a recording head and attach it to a pickup reel was too time-consuming and complicated for the vast majority of domestic users. There were some developments with cartridges for vehicle stereo systems, but it was the dictating machines used in the late 1950s that pushed advances further. Composed of two hubs and with two reels instead of one, these devices allowed recordings to be rewound and advanced forward, and several companies even used tape cassettes as portable dictation machines.

The ideas taken from these dictating machines resulted in Philips developing the cassette tape in 1962, followed closely by the compact cassette tape in 1963. The offer of licensing the technology to other companies combined with low production costs and a modest sound quality helped it become a standard format by the end of the 1960s (Nijsen, 1983). That being said, there was still an issue with the sound quality compared to vinyl due to tape hiss, but it could match larger reel-to-reel recorders in frequency response, dynamics, and distortion. To address this problem, Ray Dolby (who was working at the Ampex Corporation at the time, but went on to start the now renowned Dolby Digital) created a noise reduction system for cassette tapes. This system, known as 'Dolby Type A' (Dolby, 1967), allowed high fidelity tapes to become possible.

By the 1980s, the cassette tape was fast closing the gap with vinyl records due to its portable nature, high-fidelity, and low cost. But this race was brought to an abrupt end with an explosion into the digital age.

The digital years

Although the digital transformation of sound was first attempted by the early telephone companies to enable more messages to be transmitted per wire, and despite experiments by laboratories in the 1930s with Pulse Code Modulation (PCM), based on earlier research (Nyquist, 1928), the technology didn't become available until the early 1980s. Published in 1928, the Nyquist Theorem proposed that it is possible to convert a continuous signal into one which is discrete, and then reconstruct this back into a continuous signal without any data loss. Three decades later, Bell Labs had the first PCM transmission system in their lab, while Japanese companies experimented with the storage of PCM values on a plastic disc, before later using a laser (another Second World War invention) to read and playback the values of sound (Iwasawa and Sato, 1977). Other early digital audio techniques used helical scan video recording systems to store audio data waveforms digitised as PCM, which was then interlaced with video, thus taking advantage of the wide bandwidth and large storage capacity of the video medium.

In short, Edison's phonograph stored sound waves as a soft groove on a cylinder. Western Electrical's recorder converted sound into varying voltages of electric current, which could then be converted into a wave in a groove, or stored on a tape in which varying levels of magnetism represented the varying electrical current. Digital recording turned sound waves into pulsating electric currents that can be measured and expressed as a binary code of digits.

Although PCM was a leap forward, there were still problems relating to the storage of the data. The solution arrived in the computers of the early 1980s as they could easily store this data on hard drives in binary digits. A second solution was the digital audio disc, developed in 1979 through a working partnership between the Philips and Sony organisations. It was named the 'Compact Disc' or CD, taking its name from the earlier compact cassettes for which both companies had worked so hard to establish as a standard format (Immink, 1998). The CD was commercially introduced in 1982, sharing similarities with the original phonogram, but with a laser beam rather than a stylus, and binary code rather than grooves. Interestingly, the seemingly odd choice of a highly specific 44.1 kHz sample rate was dictated by the requirement for it to be compatible with videotape recorders used in television systems across the world at the time.

Once there were ubiquitous methods to capture and reproduce audio, the digital nature of the data made it inevitable that tools to edit audio would follow. The first digital audio workstation (or DAW) was released by Digidesign in 1989. It was entitled 'Sound Tools,' and allowed editing of the DAT (Digital Audio Tape) medium (Pope and van Rossum, 1995). By the early 1990s, home computers were being sold with built-in CD readers, and the manipulation of audio data was now becoming increasingly simple (as changing binary digits is much easier than the splicing of physical tape). It was now possible for anybody to perform basic edits at home on audio editing software such as Cool Edit Pro and other tools which would have previously only been available in high-end recording studios (Culley, 1990).

Digital data allowed an increased amount of storage, but streaming data over the newly invented internet was still an issue, due to a combination of slow transfer speeds and the relatively large file sizes of audio recordings (1 minute of WAV PCM audio required 10 MB of data). Many companies went to work on solving this problem, including Apple, who released the first version of their Quicktime format in 1991. The entire process of looking for ways to reduce file sizes was overseen by the International Organization for Standardization (ISO), who published standards for MPEG-1 in the late 1980s, followed by MPEG-2 in the early 1990s (Watkinson, 2004). At the core of the MPEG-1 audio specifications was MP3, a system developed by the Fraunhofer Institute in Germany for digital audio streaming, which at the time was just another standard and not the revolutionary codec it has since become (Musmann, 2006).

Once the internet became a necessity rather than a luxury in the late 1990s, the requirement for smaller file sizes became even more apparent. In 1997, a program named Winamp (a portmanteau of 'Windows' and 'Amp') was released by two former students of the University of Utah that could read encoded MP3 files and play them back on a home PC. This was closely followed by the first portable MP3 players in 1999 (Neal, 2013). Apple then cemented compressed audio's place in history with the release of the iPod in 2001 (Apple Computer, Inc, 2001). It was only a matter of time before dedicated portable recorders followed, and eventually, software came pre-installed on various consumer-level products, including mobile phones and smart home devices for the capture and playback of audio.

Audio forensics' beginnings

Without the aforementioned technological advances, there would be very few, if any, audio recordings used in court, and thus, no requirement for audio forensics. There are, therefore, three distinct factors which make audio forensics possible as a discipline:

1   A means to capture and playback sound.
    Digital data, and particularly compressed formats, mean every mobile phone is now capable of recording, in both standalone audio and as a stream within a video file. There are also numerous recorders (both portable and system-based) for capture in various forms, from telephone call recordings systems through to domestic doorbells with built-in surveillance. As the majority of people in the western world own mobile phones, they are, by design, carrying devices capable of the recording and playback of the sound at all times.
2   A means to edit sound.
    The editing of digital recordings is not only straightforward due to its binary format, but the software to perform such tasks is available free of charge, and in some cases even comes included as standard within recording software, enabling anybody with a minimal level of skill and understanding of audio to perform an edit. It also allows the enhancement of audio to be performed by manipulating the binary digits using digital signal processing methods.
3   A means to analyse sound.
    Since the spectrogram and the first audio forensics cases, an increasing number of analyses have become available, thanks to research performed and presented by universities, law enforcement agencies, and software development companies around the world. Although analyses were available during the analogue era, the digital age allows data to be easily, quickly, and effectively processed, and a myriad of methods exist to extract pertinent data from a recording for analysis.

Once recordings became more prevalent, so too did their use as evidence, and subsequently the need for audio forensics. Although the capture and playback of sound are of great use to the justice system by providing an objective record of sounds which occurred at a specific point in time, initially the analysis of such was limited. In terms of representation, sound was initially only visible in the time domain as a waveform, and it wasn't until the invention of the spectrogram that scientific analysis could really start to be applied to audio recordings, marking the beginnings of audio forensics as a discipline.

Although the first spectrogram was announced to the world in the November 1945 edition of *Science Journal* by Ralph. K. Potter of Bell Telephone Laboratories (Potter, 1945), they were, in fact, being used some years before this. Their development had actually begun before the Second World War, and once the war started, they obtained an official rating as a war project due to their potential as a tool for intelligence. Details of their use during the war can be traced, thanks to US Secretary of Defense documents which were declassified in 1960. According to one report, the spectrogram was first demonstrated to National Defense Research Committee (NDRC) representatives in 1941, resulting in Project C-43 being organised to provide a spectrogram to decipher speech and sounds visually, although the top priority was the study of encryption and decryption of confidential voice transmissions (National Defense Research Committee, 1944a). The type of

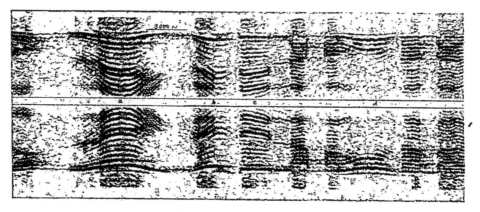

FIGURE 47. Simple inversion.

The upper spectrogram was made in the normal manner, showing inverted speech; the lower one was made with reversed oscillator sweep, producing a mechanically inverted spectrogram in which the speech appears right side up.

This sample contains harmonics with marked curvatures. These are voice inflections, and their occurrence can easily be recognized by ear. In general, samples with such voice inflections should be captured because they are most useful for diagnosing scrambling systems.

In the upper spectrogram, at points a and d, note how the curvature of the harmonics is least at the highest frequencies and progressively greater toward the lowest frequencies. Similarly, at points

b and c, the slopes of the harmonics are least at the top and greatest at the bottom of the spectrogram. This is directly the reverse of normal speech and definitely indicates inversion. The lower spectrogram illustrates the normal slopes and curvatures.

There is obviously a low-pass filter in the system, at about 3,000 cycles, as indicated by the rather abrupt change in intensity. Such a filter is normally used to cut off the upper sideband. It is usually also designed to cut off the carrier. In this case, its cutoff frequency is lower than the carrier frequency. This shows up at points a and b in the harmonics, fading out before they completely flatten out. However, the inversion frequency is not far from 3,000 cycles, because the slopes are substantially zero as they approach this frequency.

SECRET

Figure 2.3 Speech scrambling technique using inversion
Source: National Defense Research Committee, 1946.

encryption used was known as speech scrambling, something which was achieved using a variety of methods (National Defense Research Committee, 1942, 1944b, 1946[BIB-023]), including:

- *Frequency inversion*: Inverting the frequency spectrum around a centre frequency (Figure 2.3).
- *Frequency displacement*: Shifting of the frequencies upwards in a linear fashion to scramble harmonic relationships.
- *Speed wobble*: Recording a signal in a format which allows the playback speed to be increased or reduced.
- *Multiplexing*: Using multiple channels alternated along multiple source signals which alternate temporally or over frequency bands.
- *Masking*: Addition or multiplication of a coding wave to the signal.

Interestingly, areas of these reports suggest intentions of aiding the deaf, and it is also proposed as an alternative to lip reading, in which people would read spectrographs rather than lips. The system worked by recording a sound onto a loop of magnetic tape, which was then replayed repeatedly into a scanning filter which was essentially a passband which moved slowly across the frequency spectrum. The output of the scanning filter was connected to a stylus and the result traced onto a loop of electrically sensitive paper. The paper and magnetic tape loops are moved in relation to one another, so the successive scanning loops are recorded side by side, building a frequency-time-intensity image.

Once the war was over, a company called Kay Electric was licensed by Bell Labs to develop and manufacture a commercial version of the spectrogram, named the Sono-graph. The device, released in 1951, provided a tool for the visual analysis of recordings, and as such, was an obvious choice of tool for use in audio forensics (Vale, 2019).

As this development coincided with the invention of live recording systems that could be used outside of a recording studio, it was only a matter of time before a recording pertaining to a criminal event was presented in court as evidence. In 1958, the case of *United States v. McKeever* (1958) had questions relating to audio recordings at its very core. According to court documentation, the defence attempted to impeach evidence from a prosecution witness named George Ball, by claiming he was giving inconsistent evidence, based on earlier recordings made by the defendant of conversations between himself and the witness. George Ball did not recall the discussions, so was permitted by the judge to listen to the recordings on headphones during open court to refresh his memory. Once this listening session had concluded, he stated that he did in fact now remember the conversation, but he did not recognise his own voice and could not remember whether the recording was of the entire conversation. The defence argued that as he recalled the discussion, there should be no need to provide evidence relating to the authenticity of the recording. Unsurprisingly, the prosecution argued that as he did not recognise his own voice and could not recall whether it was the entire conversation, the defence should be required to provide evidence of the accuracy and authenticity of the recording. After reviewing precedents set by other cases, the judge ruled that before an audio recording could be admitted as evidence, its authenticity must be established by providing proof of the following (Maher, 2009):

1. The recording device was capable of capturing the conversation now offered in evidence.
2. The operator of the device was competent to operate the device.
3. The recording is authentic and correct.
4. Changes, additions, or deletions have not been made in the recording.
5. The recording has been preserved in a manner that is shown to the court.
6. The speakers are identified.
7. The conversation elicited was made voluntarily and in good faith, without any kind of inducement.

Eventually, a transcription of the captured conversation was determined to be admissible rather than the recording itself. Until this point, there had been no real precedent as the capture of recordings outside of a studio environment was a relatively novel occurrence. It was, therefore, this case that set out the seven specific requirements for audio exhibits to be accepted in US courts as evidence.

The new audio technology, which allowed the capture of sound in remote locations and analysis of these sounds within a laboratory, had the obvious potential for use in a courtroom. This, in turn, propelled the FBI to start developing experts in audio forensics to enhance, authenticate, and analyse audio recordings in the early 1960s (Koenig, 1990).

Notable cases

By the early 1970s, audio forensics was beginning to slowly develop as a requirement for the discipline was becoming more evident, but it wasn't until the 31st of May 1974, that it became solidified as a forensic discipline on the release of a forensic report into the EOB (Executive Office Building) tape of the 20th of June 1972 (Advisory Panel on White House Tapes, 1974). The details which led to the analysis and the subsequent report can be summarised as follows:

During the presidency of Richard Nixon, and on his command, the Secret Service installed tape recorders in several locations including the Oval Office, the Cabinet Room, the Executive Office Building, the Lincoln sitting room, and Camp David to capture telephone calls and meetings involving the President. On the 17th of June 1972, a group of five men broke into the Democratic National Committee (DNC) headquarters at the Watergate offices in Washington, DC. After an investigation, the burglars were linked to Nixon's campaign group, and further investigations led to the discovery of the secret recording devices. After mounting pressure, the tapes were released, and one in particular, made on the 20th of June 1972, was found to contain an 18½-minute gap at one point in the tape. Nixon's secretary claimed to have made a mistake by pressing the record button accidentally, wiping approximately 5 minutes of audio, but could not explain the loss of the remaining 13 minutes (Richard Nixon Museum and Library, 1977).

A forensic authentication examination of this tape with a specific focus on the 18½-minute section was requested and subsequently performed by the leading audio experts at the time, namely, Richard H. Bolt, Franklin S. Cooper, James L. Flanagan, John G. McKnight, Thomas G. Stockham Jr, and Mark R. Weiss (Figure 2.4). Many of the techniques used, or variations of such, are still in use today, and although the recording medium was analogue, some of the analysis was performed using digital computers. Although not all of the analyses made the transition from the analogue to the digital world (as they cannot be applied to digital audio), the report format used is highly relevant to this day for reporting the results of an audio recording authentication examination.

This infamous case brought the legal and audio worlds together, primarily because the level of analysis had been developed to the point that real science could now be applied to recordings. All that was required was an event to spark the audio forensics fuse, and the Watergate Scandal provided it.

Since this case, there have been a number of other high-profile cases in which audio was one of the primary pieces of evidence, but none with the impact of the 18½-minute tape. Both the murders of John F. Kennedy in 1963 (Koenig, 1983) and Trayvon Martin in 2012 (*State of Florida v. George Zimmerman*, 2013) involved evidence of the fatal gunshots, one captured by a reporter's mobile recording device and the other by an emergency service telephone system. The former was subject to an extensive audio analysis prepared by audio forensics experts in 1979 for the Select Committee on the assassination. The focus of the analysis was to determine if any potential evidence was contained within two recordings made by the Dallas Police Department on two channels of their dispatch system. The extensive analysis (Barger et al., 1979) reported that channel one was determined to have been captured 120–160 feet behind the President's limousine and contained four gunshots. Numerous FBI investigations of organised crime groups have used audio surveillance such as wiretaps and bugs, the most famous of which was in the case of New York Mafia family crime boss John Gotti (*The New York Times*, 1999). The UK

Judge John J. Sirica
United States District Court
for the District of Columbia
Washington, D.C.

Dear Judge Sirica:

We are pleased to submit herewith the final report on our
technical investigation of a tape recorded in the Executive
Office Building on June 20, 1972. This is the tape on which an
eighteen and one-half minute section of buzz appears.

The report itself occupies the first fifty pages of this volume.
The remaining pages contain appended material concerning our
study, followed by a set of detailed Technical Notes on the
scientific techniques we used and the test results we obtained.

                              Respectfully yours,

                              Richard H. Bolt
                              Richard H. Bolt

                              Franklin S. Cooper
                              Franklin S. Cooper

                              James L. Flanagan
                              James L. Flanagan

                              John G. McKnight
                              John G. McKnight

                              Thomas G. Stockham, Jr.
                              Thomas G. Stockham, Jr.

                              Mark R. Weiss
                              Mark R. Weiss

Figure 2.4 Nixon tape audio forensic report cover page
Source: Advisory Panel on White House Tapes, 1974. The EOB Tape of the 20th June 1972.

case of the Yorkshire Ripper, in which 13 murders were committed between 1975 and 1980 in the north of England also involved an audio recording. Although a caller claiming to be the perpetrator was thought to be a hoax by two audio examiners from the University of Leeds, based on voice comparison examinations, they were ignored by the police. It was later confirmed to be a hoax (French et al., 2007), which wasted valuable police resources and time. These few cases alone show the variation in the types of recordings, their content, and their application to the legal system.

Summary

Without the significant technological developments made over the last 150 years, audio forensics as a discipline would not exist. From the first capture of sound ever, through developments spurred on by the Second World War, to the landmark cases which made clear the requirement of audio analysis concerning legal matters. We now live in a time when audio forensics is more crucial than ever before due to the proliferation of audio recordings and the ease with which they can be manipulated. With the recent developments in AI manipulations, the need for this discipline is only set to increase.

References

Advisory Panel on White House Tapes, 1974. The EOB tape of June 20, 1972. Available at: www.aes.org/aeshc/docs/forensic.audio/watergate.tapes.report.pdf
Apple Computer, Inc, 2001. Apple presents iPod. Press release.
Audio Engineering Society, 2014. An audio timeline. Available at: www.aes.org/aeshc/docs/audio.history.timeline.html
Barger, J.E., Robinson, S.P., Schmidt, E.C., and Wolf, J.J., 1979. Analysis of recorded sounds relating to the assassination of President John F. Kennedy. In Appendix to Hearings Before the Select Committee on Assassinations of the House of Representatives Ninety-Fifth Congress, VIII. U.S. Government Printing Office, Washington, DC.
Bouchard, J., 2012. Older technological sound recording mediums: Problems of preservation and accessibility. State University of New York Press, New York.
Brock-Nannestad, G. and Fontaine, J., 2008. Early use of the Scott-Koenig phonautograph for documenting performance. The Journal of the Acoustical Society of America 123, 6239–6244.
Clark, M., 1993. Audio technology in the United States to 1943 and its relationship to magnetic recording. Paper presented at the 94th Convention, the Audio Engineering Society, Berlin, Germany.
Collins, T.M. and Gitelman, L., 2002. Thomas Edison and modern America: A brief history with documents. Saint Martin's Press Inc, New York.
Culley, S., 1990. Integrated audio using a desktop computer. Presented at the UK 4th AES Conference: Hard disk recording (HDR), Audio Engineering Society.
Dolby, R., 1967. An audio noise reduction system. The Journal of the Audio Engineering Society 15, 383–388.
Engel, F. and Hammar, P., 2006. A selected history of magnetic recording. Available at: www.richardhess.com/tape/history/Engel_Hammar--Magnetic_Tape_History.pdf
French, P., Harrison, P., and Lewis, J.W., 2007. R v John Samuel Humble: The Yorkshire Ripper hoaxer trial. International Journal of Speech, Language and the Law 13, 255–274.

Immink, K.A., 1998. The compact disc story. Journal of the Audio Engineering Society 46, 458–460, 462, 464, 465.

Iwasawa, T. and Sato, T., 1977. Development of the PCM laser sound disc and player. Paper presented at the 58th AES Convention, the Audio Engineering Society, New York.

Koenig, B., 1983. Acoustic gunshot analysis: The Kennedy assassination and beyond. FBI Law Enforcement Bulletin 52, 1–9.

Koenig, B.E., 1990. Authentication of forensic audio recordings. Journal of the Audio Engineering Society 38, 3–33.

Krause, M., 2002. The legendary 'Magnetophon' of AEG. Paper presented at the 112th AES Convention, the Audio Engineering Society, Munich, Germany.

Maher, R., 2009. Audio forensic examination. IEEE Signal Processing Magazine 26, 84–94.

Maraniss, H.S., 1937. A dog has nine lives: The story of the phonograph. The ANNALS of the American Academy of Political and Social Science 193, 8–13.

Mayfield, J.P. and Harrison, H.C., 1926. Methods of high quality recording and reproducing of music and speech based on telephone research. Bell System Technical Journal, 5, 493–523.

Millard, A., 2005. America on record: A history of recorded sound. Cambridge University Press, Cambridge.

Musmann, H.G., 2006. Genesis of the MP3 audio coding standard. IEEE Transactions on Consumer Electronics 52, 1043–1049.

National Defense Research Committee, 1942. Final report on Project C-55: Telegraphy applied to TDS speech secrecy system. NDRC, Washington, DC.

National Defense Research Committee, 1944a. Final report on Project C-43: Continuation of decoding speech codes. NDRC, Washington, DC.

National Defense Research Committee, 1944b. Spectrographs for field decoding work. NDRC, Washington, DC.

National Defense Research Committee, 1946. Speech and facsimile scrambling and decoding, Summary Technical Report of Division 13, NDRC, Washington, DC.

Neal, R.W., 2013. Winamp R.I.P: Celebrating the life of The Nullsoft's revolutionary MP3 player. International Business Times, November 21.

Newville, L.J., 1959. Development of the phonograph at Alexander Graham Bell's Volta laboratory (No. 5). United States National Museum Bulletin 218.

Nijsen, C.G., 1983. And the music went round on rolls, discs or reels. Paper presented at the 73rd AES Convention, the Audio Engineering Society, Eindhoven, The Netherlands.

Nyquist, H., 1928. Certain topics in telegraph transmission theory. Transactions of the American Institute of Electrical Engineering 47, 617–644.

Pope, S.T. and van Rossum, G., 1995. Machine tongues XVIII: A child's garden of sound file formats. Computer Music Journal 19.

Potter, R.K., 1945. Visible patterns of sound. Science 102, 463–470.

Poulsen, V., 1900. Method of recording and reproducing sounds or signals. 661619.

Poulsen, V. and Pederson, P.O., 1907. Telegraphone. 873 083.

Richard Nixon Museum and Library, 1977. White House tapes.

Rosen, J., 2008. Researchers play tune recorded before Edison. The New York Times. March 27.

The New York Times, 1999. Judge denies Gotti request to bar tapes from wiretap. March 17.

The New York Times, 1925. New music machine thrills all hearers at first test here. October 7.

Thiele, H.H.K., 1989. Evaluating audio inventions and innovations. Paper presented at the 86th AES Convention, the Audio Engineering Society, Hamburg, Germany.

Thiele, H.H.K., 1990. 50 years of AC bias: Dr. Walter Weber accomplishes hi-fi recording. Paper presented at the 88th AES Convention, the Audio Engineering Society, Montreux, Switzerland.

Vale, A., 2019. Sonograph. A cartooned spectral model for music composition. Paper presented at the Sound Music and Computing Conference, Malaga, Spain.

Voelker, E.-J., and Fischer, S., 2002. Philipp Reis – From the first telephone to the first microphone. Paper presented at the 112th AES Convention, the Audio Engineering Society, Munich, Germany.

Watkinson, J., 2004. The MPEG handbook. Focal Press, New York.

Wente, E.C., 1917. A condenser transmitter as a uniformly sensitive instrument for the absolute measurement of sound intensity. Physical Review 10, 39–63.

Wile, R.R., 1982. The Edison invention of the phonograph. Association for Recorded Sound Collections 14.

Chapter 3

Forensic principles

Introduction

Forensics is the application of science to legal matters, and without abiding by a set of specific principles, it would be unlikely that any work performed would be accepted by a court of law. These include those which are extremely broad but relevant to all forensic disciplines and ones which are more focused and relevant to digital forensics only. Understanding the principles is as important as the science itself, as if they are not adhered to, even the most scientific of examinations by the world's foremost authority may be deemed inadmissible.

End users

There are primarily two reasons for work to be carried out within a forensic context: (1) for investigatory purposes; and (2) for trials. Although the reasons for work are different, the processes and principles behind the work are the same.

Investigatory purposes can be considered to be those where the focus is on the identification and recovery of information (SWGDE, 2014a). These include that carried out by security services in fast-moving events where real-time seizure and monitoring are required to ensure the safety of the general public. For example, imagine a scenario in which information has been intercepted with regards to a gang planning to blow up a building. A dial-in probe is placed in the car of one of the gang members, and they are heard discussing the location of the bomb-making materials with another individual. This then allows the security services to work in real time to locate the materials and have the individuals detained at the scene. It may be that the capture from the dial-in probe is then used as evidence, in which case, it would be used for both investigatory and trial purposes, but there are numerous scenarios in which there would be no need for the audio evidence due to a confession or the availability of other overwhelmingly damming evidence.

Whether an expert is permitted to provide evidence is dependent on the system in which the trial is taking place. In less serious cases, there may be a lack of financial resources, meaning the funds just aren't available to pay for any kind of forensic work to be undertaken. The instructing of an expert may be requested and paid for privately for investigatory purposes to aid an individual's legal team, but would not be used within the court.

All generated exhibits and reports will eventually be provided to an end user, defined as the individual or group that make decisions based on the evidence. By far the most common end user is that of a judge or jury within a court system, as it is the legal system for whom the majority of work is required. As opposed to work performed for investigatory purposes, forensics for trial purposes is performed after the fact (or crime) rather than in real time.

Admissibility

Admissibility relates to the acceptance of evidence for a trial. In contrast, inadmissibility is the term for evidence being rejected based on its unreliability or it falling short of the criteria of a specific system (Cosic and Cosic, 2012). It may be that the expert cannot give evidence due to previous malpractice or lack of expertise, or that the expert is a pioneer in their field, but the evidence is unreliable due to corruption, manipulation, or being of unknown provenance. Different courts have different principles with regards to admissibility, but it is a concept that is essential within all forensic science disciplines. It is therefore of paramount importance that the rules relating to admissibility for the region in which the court is located are sought out, fully understood, and followed. As there are a myriad of systems throughout the world, a general overview will be given containing universal principles which pertain to all geographic areas and trial types.

In general, the role of an expert witness is to provide the court with expertise in a specific subject area to assist those within the justice system in understanding the evidence. The expert witness must, therefore, be able to demonstrate that the information they are providing is reliable, by being able to evidence a level of expertise in their specialist area, whether through their qualifications, research, training, or experience (Smith and Bace, 2002). There have been well-documented instances when trials have fallen apart when it has come to light that a key expert witness has no qualifications or experience in the field for which they have given evidence. As experts are independent of the court system, any history of bias towards one party or another, whether through manipulation of evidence or by purposely providing inaccurate or incorrect information will likely deem the expert's testimony inadmissible.

In terms of evidence recordings, if questions are raised over the authenticity, the burden is generally on the party raising the concern to provide evidence as to the reasons for this. It may then necessitate a separate 'evidentiary hearing' focused on this issue before the trial in chief takes place. Evidence which is shown to be an inaccurate representation of the original should then be ruled as inadmissible. The Audio Engineering Society (Audio Engineering Society, 2012) defines an authentic recording as that which is:

> made simultaneously with the acoustic events it purports to have recorded, made in a manner fully and completely consistent with the methods of recording claimed by the party who produced the recording and free from unexplained artifacts, alterations, additions, deletions or edits.

If it is shown that the recording used for analysis is not a bit-stream copy of the original recording, meaning it is not an exact bit-for-bit copy, it should be inadmissible. The reasons for this are many, including the fact that it may be that the recording analysed is of inferior quality to the original version, leading to reduced intelligibility, or at worst, misinterpretation of the content. Another reason for inadmissibility may be the trimming of a recording to contain only a specific section, causing 'framing bias' in which only the material that is favourable to one party's case is presented, and the rest removed. This can lead to a change in both semantics and context, which could have a real impact on how a recording is interpreted. Manipulation in any form is an obvious reason for inadmissibility as the recording would be entirely unreliable, and subsequently, any enhancement or analysis of such would also be inadmissible.

Although the list of reasons for inadmissibility seems extensive, it is not difficult to ensure evidence is admissible if the basic principles are adhered to, and company procedures are followed. Various organisations, from both the scientific and legal sectors, produce best practice guidelines to aid examiners in preventing any issues relating to such.

Best practices

Best practice documents are published by departments in government, committees in international standards organisations, and scientific working groups (groups of scientists from a specific field). They are often publicly available and encompass guidance, standards, and procedures pertaining to the best way to practise forensic science.

By having access to these guidelines, all forensic scientists working within a field have a point of reference to ensure the aforementioned admissibility issue is negated, there is a standard that examiners are expected to meet, and scientific principles are upheld. As it applies to audio, there are best practice guidelines which are both general and specific. The groups that publish the documents are usually experts in the field who collaborate to create guidelines over an extended period which can be several years in some cases. A draft form of the document is then made public to allow other scientists in the field to provide feedback. Considerations are subsequently made with regards to these responses, and any necessary amendments are performed before the document is officially published. The documents are then periodically reviewed and updated when required, which, again, will be released for comment before being officially published.

In terms of the organisations that produce standards, there are several of note, and there is, as one might expect, some degree of overlap between the content of the documents. The youngest of these groups is the Scientific Working Group for Digital Evidence (SWGDE), which was first formed by the Federal Crime Laboratory Directors in 1998 to address the growing volume of digital evidence in the form of computers, mobile phones, tablets, and multimedia. This spawned several sub-committees, including one for audio forensics. It started as an ad-hoc group of digital forensic examiners in 2005, before becoming a standing committee in 2006 (Pollitt, 2003a). Their first publication with regards to audio was entitled 'Best Practices for Forensic Audio' (SWGDE, 2008), intended to provide recommendations and advice on the handling and examination of audio evidence to ensure it is suitable for use within the context of the legal system. It includes information on laboratory setup, evidence handling, examination preparation, delivery of results, and general guidelines for the duplication, repair, and enhancement of audio recordings. Since that first document, a number of others have followed, including:

- SWGDE, 'Electric Network Frequency Discussion Paper' (SWGDE, 2014b);
- SWGDE, 'Core Competencies for Forensic Audio' (SWGDE, 2017);
- SWDGE, 'Best Practices for Digital Evidence Collection' (SWGDE, 2018a);

- SWGDE, 'Best Practices for Digital Audio Authentication' (SWGDE, 2018b);
- SWGDE, 'Video and Audio Redaction Guidelines' (SWGDE, 2018c).

A second organisation, the European Network of Forensic Science Institutes or ENFSI for short (the most extensive European forensic science working group), is composed of numerous field-specific subgroups, including one for audio forensics. Membership requires being part of a European organisation which has a minimum number of employees and is working towards ISO 17025 accreditation (Kjeldsen and Neuteboom, 2015). ENFSI's all-encompassing digital forensics best practice guideline is called 'Guidelines for Best Practice in Forensic Examination of Digital Technology' (ENFSI, 2015) and aims to provide a framework of standards, quality principles, and approaches in compliance with ISO 17025. It was first agreed in 2003 by the ENFSI IT working group and has since undergone several revisions to keep abreast with the latest research and technology.

The 'ENFSI Expert Working Group Forensic Speech and Audio Analysis (FSAAWG)' is the title of the audio Working Group and is composed of members from 20 countries. Their only publication at the time of writing is 'Best Practice Guidelines for ENF Analysis in Forensic Authentication of Digital Evidence' (ENFSI, 2009).

In the UK, the main body producing guidelines is the Forensic Science Regulator, or FSR, who have a strong focus on compliance with ISO 17025. Their only publication to date concerning audio forensics is 'Speech and Audio Forensics Services' (Forensic Science Regulator, 2016). Rather than have specific working groups, they write the best practice documents with support from technical experts in each field.

The most audio-specific group is the Audio Engineering Society (AES), who too have their own technical committee in the field of audio forensics. They are also a prolific publisher of peer-reviewed audio forensics-related research. In 1996, a document was developed by the Working Group on Forensic Audio of the Audio Engineering Society Standards Committee, entitled 'Managing Recorded Audio Materials Intended for Examination' (Audio Engineering Society, 1996) and has undergone several revisions since. The Working Group was formed in 1991 at the request of a community of engineers within the AES, the Acoustical Society of America, various law enforcement agencies, and other groups who were concerned with standards of audio forensics-related expert testimony. The methodologies included are based on those first developed within the 'Report on a Technical Investigation Conducted for the US District Court for the District of Columbia' by the Advisory Panel on the White House Tapes covered in Chapter 2 (Advisory Panel on White House Tapes, 1974).

One further organisation, ASTM International (American Society for Testing and Materials) has various standards which are for sale rather than provided free of charge. The only one that is directly related to audio forensics is ASTM E3150-18, or the Standard Guide for Forensic Audio Laboratory Setup and Maintenance (ASTM International, 2018), which documents recommendations for the design, configuration, verification, and maintenance of audio forensics laboratories.

Certain best practices relate to all digital evidence, regardless of the type of evidence at the heart of a specific discipline. Although they are usually adopted and reworded by various organisations, the fundamental principles are the same in each document. The first set of principles were created in the early 1990s by the International Association of Counter Investigative Specialists (IACIS). Today, the most commonly applicable standard was that created by the IOCE (International Organisation for Computer Evidence) at the Hi-Tech Crime and Forensic Conference (IHCFC) which took place on 4–7 October 1999 (Pollitt, 2003b). During the meeting, the UK Association of Chief Police Officers Good Practice Guide for Digital Evidence (ACPO, 2012) and several draft standards from other organisations were reviewed. The result was unanimous agreement by the delegates on the following set of principles:

1   Upon seizing digital evidence, actions taken should not change that evidence.
2   When it is necessary for a person to access original digital evidence, that person must be forensically competent.
3   All activity relating to the seizure, access, storage, or transfer of digital evidence must be fully documented, preserved, and available to review.
4   An individual is responsible for all actions taken concerning digital evidence while the digital evidence in their possession.
5   Any agency that is responsible for seizing, accessing, storing, or transferring digital evidence is responsible for compliance with these principles. (FBI, 2000)

Although there is already an extensive array of best practice guidelines available, the scope increases year on year with the publication of further documents. Those that are already published are regularly updated as they are intended to be living documents that are consistently reviewed and amended, based on advances in research, technology, and laws. There is a degree of overlap, but it is worth reading all the relevant guidelines as some cover certain topics in greater depth than others. It is recommended that the reader build their own digital library of the said guidelines, retiring and updating documents as new versions are published, thus ensuring they are readily available for review and the most current best practices are always adhered to.

The importance of following best practice guidelines cannot be understated, as in doing so, all potential issues previously discussed in relation to admissibility are negated. By also implementing the other practices specific to audio, and introducing elements, such as standard operating procedures and peer review, based on the said guidelines, the reliability of results can only be improved.

Standard operating procedures

Standard operating procedures (or SOPs as they are more commonly called) are documents created by laboratories detailing specific methodologies to ensure the scientific principles of repeatability and reproducibility are met. Following SOPs ensures error mitigation, the reduced liability of examiners' actions, consistency, and quality assurance within the forensic laboratory (Slay et al., 2009).

Although there are procedures available online, they are generally organisation-specific and written by examiners who perform the tasks within their labs, but do follow the aforementioned guidelines (SWGDE, 2012). An SOP document is analogous to a recipe, in which the 'recipe' itself is the technique, for example, 'Exhibit Imaging.' The ingredients are the tools required to perform the task, so in the case of imaging, it may include an external device containing the file (such as a portable audio recorder) and a write-blocker (to ensure data on the device remains unchanged when connected to a forensic workstation). The procedure itself is akin to the instructions for the recipe itself, provided in a level of detail that would allow another examiner trained to a level of competency set out within the document to perform the same procedure and achieve the same result.

All standard operating procedures should also include the scope and limitations of the method to prevent it from being applied to unsuitable data or using tools which have not been validated and may cause unpredictable results.

The implementation of standard operating procedures should be an iterative process, in which new guidelines, tools, and techniques are introduced when and where it is deemed necessary to ensure the methods are up to date and are as efficient and reliable as possible (Meyers and Rogers, 2004). They should also be reviewed after any system updates or unexpected results which may indicate there are errors with the method or that certain data types are outside of their scope. Once updated, the previous version must be retired so there is only one SOP in circulation for a specific task at any one time. Past versions should be kept on file to ensure they are available so any questions raised over historical cases for which the procedure was applied can be answered accurately.

Bias mitigation

You must not fool yourself, and you are the easiest person to fool.

(Fenyman, 1964)

Cognitive biases pose a risk to both science and the justice system, so it is no surprise that they are an issue that needs addressing and measures should be put in place to mitigate against their influence. Biases can be introduced at any stage, and are often very subtle, meaning they have the potential to go unnoticed by both the examiner and be overlooked during a peer review. For this reason, steps must be put in place from the initial contact with a potential client through to giving evidence in court to ensure results and conclusions are as objective as possible and based solely on findings that can be supported from the evidence reviewed. The following section provides an overview of the principal types of bias of which all forensic audio examiners should be aware.

Confirmation bias

Kassin et al. (2013) define the term 'forensic confirmation bias' as 'the class of effects through which an individual's pre-existing beliefs, expectations, motives and situational content influence the collection, perception and interpretation of evidence during the course of a criminal case.' This can occur either consciously or subconsciously, and often originates from extraneous information being provided to the examiner which is outside the scope of that required to complete their work. For example, imagine an examiner is performing an authentication examination of an audio recording and is told the suspect works in a recording studio. They may then wrongly assign a higher weighting to certain findings which otherwise would not influence the conclusion. The expectation effect in which an examiner 'expects' an outcome contributes to confirmation bias, especially so when there is too much information for consideration, leading one to selectively review the evidence and focus on specific pieces rather than the evidence in its entirety (Saks et al., 2003).

Another form of confirmation bias relates to the base rate (or the expected norm). A prime example is a peer reviewer who, over time, begins to expect to agree with the findings of another examiner's report if they rarely disagree (and thus assume the base rate). One proposed solution is to perform a blind review, where only the forensic decisions made during the work are reviewed, and the final conclusion is kept hidden. Another is to create dummy reports on a regular basis (unbeknownst to the reviewer who believes it to be a real case) containing data that opposes the base rate. This is a technique performed at airport security by including dummy cases containing fake bombs and is designed to keep reviewers on their toes (Dror, 2013).

Base-rate fallacy

The base-rate fallacy is defined as 'people's tendency to ignore base rates in favour of individuating information rather than integrating the two' (Bar-Hillel, 1980). This relates to an individual prioritising new information (for example, information pertaining to a specific case) over pre-existing, or base rate information (general, pre-existing information which has been shown to be correct time after time). An understanding of this is essential when presenting or reporting findings to ensure they are not misinterpreted by the trier of facts, as base rate information can be highly relevant when drawing conclusions. Base rate information is also the basis for Bayesian analysis, a statistical approach named after Thomas Bayes that explains how probabilities change in light of new evidence (Neal and Grisso, 2014).

Prosecutor's fallacy

The previously mentioned base rate fallacy is a form of prosecutor's fallacy, one of the most common, yet misunderstood, forms of bias. According to Thompson (Thompson and Schumann, 1987), its foundations lie in the 'confusion about the implications of conditional probabilities.' It is also known as the transposed conditional, where the prior odds (before considering other conditions) are ignored, and likelihood ratios (the probability of the *evidence* given a *hypothesis*) are misinterpreted as the posterior odds (the probability of the *hypothesis* given the *evidence*). An example given within the referenced paper can be surmised as thus: All lawyers own briefcases, but only one person in ten in the general population owns a briefcase (the posterior odds). When asked to judge the probability a man is a lawyer as he owns a briefcase, the prosecutor's fallacy would be to conclude there is a 90 per cent chance the man is a lawyer. We know that there are more briefcase owners in the general population who are not lawyers (the prior odds), and as such, there is a higher probability that the owner of the briefcase is not a lawyer. When applied to forensics, it is the process of drawing incorrect conclusions about the likelihood that a suspect is guilty, based on the evidence of a 'match' without considering the *a priori* likelihood based on the percentage of other people who would also match.

Texas sharpshooter's fallacy

This is the term for an informal fallacy given by epidemiologists to define the act of giving significance to random data after an event has occurred. It takes its name from the story of a Texan who fired a rifle into a barn before painting around each of the bullet holes to falsely demonstrate the perceived accuracy of the shot. It is also known as painting the target around the arrow (Thompson, 2009).

An example in audio forensics would be adjusting the criteria for a match within a speaker comparison examination after the analysis has taken place to improve results (in effect, moving the target to match the outcome). To a third party, this may appear to improve the accuracy of the analysis, while it has, in fact, reduced it, as the criteria required to result in a match have been broadened.

Optimism bias

Optimism bias is defined by Sharot (2011) as 'the difference between a person's expectation and the outcome that follows,' and can be caused by an examiner's optimism in the methodologies and tools used. For example, applying a noise reduction process to an audio recording in the belief that the overall quality has improved when it has, in fact, reduced the intelligibility of the speech.

Contextual bias

Contextual bias is described by Venville (2011) as events in which 'well-intended experts are vulnerable to making erroneous decisions by extraneous influences.' An obvious example would be knowing the conclusions of other investigations relating to a case, such as the results of a DNA test relating to a subject for whom a voice comparison exercise has been requested.

Framing bias is a form of contextual bias where an examiner only analyses a section of the information available (hence 'framing'), and draws their conclusions from such without consideration for the rest of the data. This not only provides a fraction of the whole picture but can also result in information being misinterpreted due to a lack of context, for example, the drawing of conclusions after performing only a couple of brief analyses during an authentication examination (Tversky and Kahneman, 1981).

Statistical bias is yet a further type of contextual bias, in which the method causes errors due to the data analysed, the analysis procedure, or the interpretation of the analysis (Rumsey, 2016). This can occur if an analysis procedure which is not fit for purpose is used, such as using a method to analyse the compression history of a file for which it is not designed, or using an incorrect reference population when performing a voice comparison examination.

Allegiance bias/adversarial allegiance

Allegiance bias is described as the presumed tendency for experts to reach conclusions that support the party who retained them, and as such, they are not, in fact, independent (Murrie et al., 2013). Theories for this range from experts beginning to think of themselves as being on a specific team (Brodsky, 2012), unintentional and known cognitive errors such as confirmation biases (Neal and Grisso, 2014), through to intentional processes and motives for financial gain (Hagen, 1997).

Biases do not exist in vacuums, and so there is a degree of cause and effect between different types, for instance, contextual bias can lead to confirmation bias. As the majority of biases are subconscious, the best mitigation technique is to limit the requirement for decisions to be made by the examiner, thus reducing the influence of any unconscious thoughts. The most obvious way to do this is by implementing procedures which are strictly adhered to, and that include specific steps to mitigate biases. When creating procedures, best practice guidelines should be followed, peer review employed, and stages included to only allow examiners access to the information required to perform the work. Instructing parties have a habit of providing examiners with all of the information they have on the case as they do not know what the examiner requires. It is not uncommon to be provided with the entire case bundle, including everything from interviews through to the defendant's criminal records. One method to prevent any biases which may arise from this is to provide the instructing party with a list of the exhibits and information required beforehand and request that nothing else is provided unless requested. A disclaimer can also be included when providing quotations to make it clear that any extraneous data that was not requested will be deleted or destroyed (assuming they are sent electronically and are therefore not original copies). Another, as proposed by Dror et al. (2015), is a system entitled Linear Sequential Unmasking. By splitting the work process into distinct stages and then only making available the information required by an examiner at each stage, the potential for some of the biases is minimised. This can easily be achieved by having a case manager or another examiner who is not working on the case act as a filter between the instructing party and the examiner, thus ensuring the examiner is only provided with that which is required. If an opinion on other experts' conclusions is requested, their report should only be reviewed once an independent analysis of the evidence has been completed, thus removing the possibility of the base-rate fallacy bias. Another method to reduce bias is to consider the opposite, an approach which forces the examiner to evaluate the evidence against opposing propositions. When this exercise is performed, it can ensure all possibilities are considered, as depending on how information is framed, conclusions can sometimes seem plausible solely due to a lack of competition (Koehler, 1991).

When reporting results, all background data and experience used to draw conclusions should be documented, and any gaps (such as information that was not provided but would have aided the examination) should be reported (Forensic Science Regulator, 2015).

Peer review

Peer review is well known within academia as the process of another suitably qualified expert (but usually more than one) reviewing a research article before it is published, thus ensuring the methodologies and findings are scientifically sound. Within forensics, this should be conducted for each case completed, but rather than being performed blindly by an individual unknown to the author (as is often the case in research), it is generally performed by somebody with a suitable level of competency within or outside of the company, or both. Having work peer-reviewed by an examiner from an external organisation is preferable as there may be subconscious allegiance biases if both the reviewer and examiner work for the same organisation, resulting in a less critical review. There are various areas of review which look at different aspects of the work, for example, a critical review of methods used, the accuracy of technical information, formatting, and grammar. The intended audience should also be taken into account, so potential areas of misinterpretation can be clarified, and jargon removed (ENFSI, 2015).

It has been shown in various research studies that the higher the number of examiners who review the work, the better the accuracy of results (Phillips et al., 2018). Having an internal non-technical peer review to pick up on issues which relate to the organisation, such as formatting, followed by an external technical review would be the ideal scenario.

The forensic laboratory set-up

The optimum design of a forensic audio laboratory is based on two core requirements: (1) the mitigation of issues which may reduce the accuracy, reliability, and objective nature of the work; and (2) the need for various audio-related tools to allow the laboratory to perform a range of examinations. The result of these criteria is a hybrid between a classic forensic laboratory (such as that used for DNA and blood analysis) and a recording studio. The following sub-sections provide further information with regards to a number of considerations which should be made when creating or maintaining a forensic audio laboratory.

Environment

The first area to be addressed is the workspace. In terms of location, the mitigation of noise and the security of evidence should be scrutinised as both could have potentially dire consequences if not addressed correctly. Forensic recordings by their nature are often noisy due to poor capture conditions, so substantial efforts should be made not to introduce further noise. As the medium which is being examined is sound, any extraneous noise must be avoided, so an ideal space would be away from sources of noise, such as main roads, HVAC (heating, ventilation, air conditioning) systems or shared offices (Koenig et al., 2007). Recording studios go to great lengths to achieve this through expensive installations using absorbent materials to soundproof the listening area. While forensic organisations may not have the money of major record labels, it could be argued that the end product for them is more important as justice may depend on it.

There are also several requirements based on equipment and health and safety. The area should be free from any form of damp or moisture, a leading cause of destruction to digital devices. Hardware containing magnets, such as speakers and microphones, should also be avoided where possible as they have the potential to wipe data from devices. If they are used, they should not be stored within the same space as evidence. Power is an obvious requirement for the forensic workstation, and a minimum of eight outlets should meet all requirements. In terms of health and safety, considerations should be made to ensure there are marked fire exits, and all electronic devices are tested in accordance with the laws of the country in which the laboratory is located.

The physical security of the area should also be a priority due to the nature of forensics and the requirement to control exhibits (Koenig, 1988). All doors into the laboratory should be fitted with access control systems that automatically store logs relating to the people entering and leaving the space. A visitor log should also be maintained to serve as a paper trail of the date and identity of visitors from outside the organisation. As access is also possible through the internet, digital security should also be taken seriously, and firewalls, regular password changes, and virus scans should be employed.

Equipment

Workstation

At the heart of all audio forensic laboratories are the computers, or 'workstations.' As computers are becoming ever more powerful, and today's high specification computers will become tomorrow's standard models, it would be of little use to document the minimum specifications of a workstation for the purposes of digital audio forensics. Instead, the recommended specifications of software to be installed on the system must be reviewed, and considerations made as to the amount of memory expected for the approximately three months' storage of casework files. Cases older than this can be removed from the system and archived to back-up storage as it is likely that they will have been completed by this point. The selection of monitors and their number are based on an individual's preference, as although some visual analysis is needed, the requirement of an extremely high definition monitor is not essential. The workstation should also be isolated from any networks to prevent automatic updates which could cause system instability or unpredictable results. A further advantage of having an air gap such from networks is that confidential casework data cannot be compromised through hacking. In recent years, this has become increasingly important due to the widespread and well-publicised breaches of various organisations' IT infrastructure, including those that work in the forensic sectors. Rather than download exhibits provided over a network directly to the workstation, another computer should be used for this task, and the exhibits downloaded directly to a password-protected and encrypted USB flash drive that is not permitted to leave the laboratory.

Software

As there are many different programs available which perform the same operations, the developers and names of software are not important. What is essential is that the examiner feels comfortable with their choice of software, that it is suitable for the task required of it, and it has been validated to ensure it works as expected. The following is a list of software considered essential for the topics covered within this book:

1   Hex editor/viewer. Used for viewing the file structure of audio files.
2   Audio enhancement software. The software should feature processing tools such a de-clip, de-click, de-reverb, de-noise, de-hum, normalisation, level adjustments.
3   Word processor. Essential for all documentation and writing of reports.
4   Case management system. Required for keeping track of the status of cases, logging exhibits, and for general case management.
5   Metadata reader. To extract and view all metadata contained within an audio file.
6   Audio conversion software. Conversion (or extraction from video) of audio to an uncompressed WAV PCM format for enhancement.
7   Audio analysis software. Software which features, at minimum, a waveform viewer, spectrogram, and FFT plot function.

The benefit of having the latest software is of less importance than the stability and predictability of results. Before updates are performed, a review of the reason for the update should be performed, and verification testing completed to ensure the software functions as expected before its application to casework.

A list of all software versions and the dates of updates must also be documented to ensure only those which have been validated and verified are used, and periodic verification is performed based on the dates stored within the log.

Audio equipment

As with computers, audio equipment is continually improving, so recommendations relating to specific specifications would be of little use as they would quickly become outdated. With that being said, there are two essential pieces of hardware equipment for digital audio forensics: (1) a set of high quality, closed-back, over-ear headphones, and (2) a sound interface with excellent digital to analogue convertors.

First, the headphones. Closed-back, over-the-ear designs trap the sound output from the headphones, prevent external sounds from entering the ear canal, and prevent reverberation, all of which reduce the possibility of any misrepresentation of the audio output due to the acoustics of the listening environment. This is essential for ensuring tasks such as critical listening and transcribing can be performed in optimal conditions. A built-in limiter would also be desirable to prevent loud transients which have the potential to damage hearing, and the frequency response should be as flat as possible to avoid regions being attenuated and boosted which may colour the sound (Bergfeld and Junte, 2017). It is better to control the frequency response by using a pair with a flat frequency response and apply boosts and cuts from within the workstation than listen to a signal which has a frequency response coloured by the speakers. Most headphones designed for commercial use by the general public will not have a flat frequency response as their primary task is to sound 'good,' which means their representation of the material is not necessarily accurate.

Second, a sound interface with a number of inputs and high-quality ADC and DAC converters. Conversion to the digital domain is required in instances where there is no digital access to recordings stored on a device, and so an analogue signal must be sent to an input on the interface. This is becoming less of a necessity as devices with various digital interfacing options, such as removable memory cards, wi-fi connections, and direct cable transfer become more common. Capture through analogue means is not ideal, but does happen on occasion. The interface similarly requires a high-quality DAC for output to the headphones.

Other devices

In addition to those previously discussed, there are a number of other devices which make up the digital aspect of a forensic audio laboratory. Not all are required, and some exist in both the hardware and software domains (for example, write blockers), so considerations should be made as to which are needed based on the type of work performed, and whether software or hardware versions are preferred. The following is a general overview of these devices:

1   Hardware write-blocker. Used to ensure evidence on original devices remains unchanged when being imaged to a workstation.
2   Portable audio recorders. Essential for the creation of reference recordings relating to casework.
3   Cassette tape player. Although not crucial, they can be required when the conversion from an analogue tape is needed.
4   External hard drives. For backing up data, a minimum of two hard drives per back-up should be used to ensure that if one is damaged, the data remains stored on the other.
5   Printer. For printing reports and other documentation in preparation for courtroom testimony.
6   DVD reader/writer. For reading exhibits provided in a CD/DVD format and burning enhancements to data DVDs where required.
7   SD card reader. For reading exhibits provided in an SD card format.

Other equipment

Although not directly related to examinations, there are other several essential pieces of equipment which may be required within the laboratory environment, depending on the tasks performed. For example:

1   Evidence bags. For the safe storage and logging of incoming physical evidence in tamper-proof bags.
2   Secure storage. A protected area of storage, such as a locked room or a safe with restricted access to ensure the safe-keeping of evidence.
3   Data DVDs. For providing enhanced audio to clients where requested.

Library

A library is an integral aspect of any forensic laboratory as it can be useful for finding information relating to tasks which are non-routine and further knowledge is sought, as well as providing references for casework. It can also be used to source reliable information for any research undertaken. It may include books covering topics that range from audio principles through to programming.

As the majority of guidelines and scientific papers are now in digital formats, a digital library is another necessity. It also has the advantage of being simple to organise and makes searching for specific terms incredibly easy. Inclusion of user manuals, research papers, and best practice guidelines is recommended.

Documentation

The requirement for documentation stems from two factors which are not only a critical element of forensics but of science in general. All scientific methodologies should be both repeatable and reproducible. Repeatability pertains to the same scientist repeating the test and obtaining the same results (within a pre-defined error-rate). In contrast, reproducibility relates to a different scientist repeating the same method as the first scientist and achieving the same results, again within a pre-defined error rate. This is a fundamental principle of science, and without this predictability, no progress would be possible as a scientist conducting new research could not rely on the research of others so would have to begin by inventing the wheel. Never has the phrase 'standing on the shoulders of giants' been more applicable than it is to scientific research.

With regards to forensic science, this can be taken a stage further. Research is essential to the creation of new techniques and methodologies, but the repeatability and reproducibility factors also apply to the methods used by scientists to ensure the justice system is provided with one further 'r,' that of reliability.

Repeatability x Reducibility = Reliability.

For these reasons, contemporaneous working notes must be taken from the moment the instruction is received through to the results of the work being delivered to the instructing party (Forensic Science Regulator, 2016). These may be in the form of exhibit logs, word-processed documents, or screenshots. They should not only contain the exhibits reviewed and the processes performed, but also the software used, the reason for any decisions made in terms of the procedures applied, and any reasoning behind opinions given. This is vital for several distinct purposes:

1   When a report is served to an opposing expert, it allows them to review the work undertaken and opine on whether they agree or disagree with the methodology and any conclusions reached.
2   If further action is required, it aids the examiner as they can review notes and pick up where they left off.
3   If additional work is required and the examiner who performed the work is unavailable, a second examiner can review the primary examiner's notes before continuing the work.
4   More often than not, the actual forensic work will be performed months in advance of a trial. Recalling the processes performed while giving evidence would be both stressful and unreliable if notes have not been taken at each stage.
5   If an opposing expert uses the same tools but obtains different results, the reasons for such can be easily determined (for example, different versions of the same software, different versions of a recording).

The sequence of documentation can be defined as follows:

1   Exhibit log
    Documentation concerning the exhibit received, regardless of whether it is a digital or physical exhibit. This may include but is not limited to: case reference, date received, method of delivery, format (digital/physical), filenames, and hash checksum of data.
2   Working notes
    These should consist of, at minimum, the following: case reference, examiner, date of analysis, results of any preliminary analysis, proposed steps, software versions used (including the

operating system), the hardware used, processing applied, any opinions, and the set of premises on which these opinions are based.

3   Peer review

Once an examiner has completed a report relating to an examination, it is peer-reviewed by another expert trained to a specific level of competency. Notes are then made by the reviewer, either digitally or physically, and provided for the examiner to consider before amending the original report where necessary. These considerations will range from terminology and grammar, the analysis performed, conclusions reached, and any legal necessities in relation to the work performed.

Although this may seem like a lot of documentation, there are methods to reduce it while still ensuring the same level of detail and thus, reliability. One example with regards to audio enhancements is to export the settings and processes applied from the software used. Not all programs include this feature, but if available, it is an option as it is simple and effective in ensuring the processes applied are documented accurately.

Summary

Forensics is built on a foundation of principles, which, if they didn't exist, would result in a discipline where each examiner did as they pleased without any form of standardisation, and thus no reproducibility. It would also be challenging to hold examiners accountable for their actions. Best practice guidelines are critical in ensuring all areas of forensics are performed to a specific standard across the board, and errors are minimised, from the laboratory environment through to the delivery of the final report. Although there appears to be a large number of guidelines for forensics, many of which contain similar and non-conflicting information, there are also some which are highly specific to certain tasks, such as audio authentication or enhancement. If the most recent best practice guidelines are adhered to, it is very difficult to drift off course, and just as difficult for work to be challenged in the courtroom under cross-examination or by other experts.

References

ACPO, 2012. ACPO good practice guide for digital evidence. ACPO, London.

Advisory Panel on White House Tapes, 1974. The EOB tape of June 20, 1972. Available at: www.aes.org/aeshc/docs/forensic.audio/watergate.tapes.report.pdf

ASTM International, 2018. Standard guide for forensic audio laboratory setup and maintenance (No. ATSM E3150-18). Available at: www.astm.org/Standards/E3150.htm

Audio Engineering Society, 1996. AES recommended practice for forensic purposes; Managing recorded audio materials intended for examination (No. AES27-1996 (s2012). Available at: www.aes.org/publications/standards/search.cfm?docID=29

Audio Engineering Society, 2012. AES recommended practice for forensic purposes: Managing recorded audio materials intended for examination. Available at: www.aes.org/publications/standards/search.cfm?docID=29

Bar-Hillel, M., 1980. The base-rate fallacy in probability judgments. Acta Psychologica, 44(3), 211–233.

Bergfeld, D. and Junte, K., 2017. The effects of peripheral stimuli and equipment used on speech intelligibility in noise. Paper presented at the 2017 AES International Conference on Audio Forensics. Audio Engineering Society.

Brodsky, S.L., 2012. Testifying in court: Guidelines for the expert witness, 2nd edn. American Psychological Association, Washington, DC.

Cosic, J. and Cosic, Z., 2012. Chain of custody and life cycle of digital evidence. Computer Technology and Application, 3, 126–129.

Dror, I.E., 2013. Practical solutions to cognitive and human factor challenges in forensic science. Forensic Science Policy & Management: An International Journal 4, 105–113.

Dror, I.E., Thompson, W.C., Meissner, C.A., Kornfield, I., Krane, D., ... Risinger, M., 2015. Letter to the Editor, Context management toolbox: A Linear Sequential Unmasking (LSU) approach for minimizing cognitive bias in forensic decision making. Journal of Forensic Sciences 60, 1111–1112.

ENFSI, 2015. Guidelines for best practice in the forensic examination of digital technology. Available at: citeseerx.ist.psu.edu/viewdoc/download?doi=10.1.1.651.252&rep=rep1&type=pdf

ENFSI, 2009. Best practice guidelines for ENF analysis in forensic authentication of digital evidence. Available at: enfsi.eu/wp-content/uploads/2016/09/forensic_speech_and_audio_analysis_wg_-_best..

FBI, 2000. Digital evidence: Standards and principles, by SWGDE and IOCE. Forensic Science Communications 2(2). Available at: www.fbi.gov/about-us/lab/forensic-science-communications/fsc/april2000/swgde.htm

Fenyman, R., 1964. What is and what should be the role of scientific culture in modern society? Lecture at the Galileo Symposium, Italy.

Forensic Science Regulator, 2015. Cognitive bias effects relevant to forensic science examinations. Available at: www.gov.uk/.../cognitive-bias-effects-relevant-to-forensic-science-examinations

Forensic Science Regulator, 2016. Speech and audio forensic services. Appendix: Available at: www.gov.uk/government/publications/speech-and-audio-forensic-servicesHagen, M.A., 1997. Whores of the court: The fraud of psychiatric testimony and the rape of American justice, 1st ed. Regan Books, New York.

Kassin, S.M., Dror, I.E., and Kukucka, J., 2013. The forensic confirmation bias: Problems, perspectives, and proposed solutions. Journal of Applied Research in Memory and Cognition 2, 42–52.

Kjeldsen, W.T. and Neuteboom, W., 2015. 20 years of forensic cooperation in Europe. ILAC News 47.

Koehler, D.J. 1991. Explanation, imagination, and confidence in judgement. Psychological Bulletin 110, 499–519.

Koenig, B.E., 1988. Enhancement of forensic audio recordings. Journal of the Audio Engineering Society 36, 884–894.

Koenig, B.E., Killion, S.A., and Lacey, D., 2007. Forensic enhancement of digital audio recordings. Journal of the Audio Engineering Society 55.

Meyers, M. and Rogers, M., 2004. Computer forensics: The need for standardization and certification. International Journal of Digital Evidence 3, 11.

Murrie, D.C., Boccaccini, M.T., Guarnera, L.A., and Rufino, K.A., 2013. Are forensic experts biased by the side that retained them? Psychological Science 24, 1889–1897.

Neal, T.M.S. and Grisso, T., 2014. The cognitive underpinnings of bias in forensic mental health evaluations. Psychology, Public Policy, and Law 20, 200–211.

Phillips, P.J., Yates, A.N., Hu, Y., Hahn, C.A., ... O'Toole, A.J., 2018. Face recognition accuracy of forensic examiners, super recognizers, and face recognition algorithms. Proceedings of the National Academy of Sciences of the U.S.A. 115, 6171–6176.

Pollitt, M.M, 2003a. Who is SWGDE and what is the history? SWGDE, 3.

Pollitt, M.M., 2003b. The very brief history of digital evidence standards. In M. Gertz (ed.), Integrity and internal control in information systems. Springer, Boston, MA, pp. 137–143. https://doi.org/10.1007/978-0-387-35693-8_8

Rumsey, D.J., 2016. Statistics for dummies, 2nd ed. Wiley, New York.

Saks, M.J., Risinger, D.M., Rosenthal, R., and Thompson, W.C., 2003. Context effects in forensic science: A review and application of the science of science to crime laboratory practice in the United States. Science & Justice 43, 77–90.

Sharot, T., 2011. The optimism bias. Current Biology 21, R941–R945.

Slay, J., Lin, Y.-C., Turnbull, B., Beckett, J., and Lin, P., 2009. Towards a formalization of digital forensics. In G. Peterson and S. Shenoi (eds), Advances in digital forensics V. Springer, Berlin, pp. 37–47.

Smith, F.C. and Bace, R.G., 2002. A guide to forensic testimony: The art and practice of presenting testimony as an expert technical witness. Addison-Wesley, Reading, MA.

SWGDE, 2008. Best practices for forensic audio, 21. Available at: www.swgde.org/documents/swgde2008/SWGDEBestPracticesforForensicAudioV1.0.pdf

SWGDE, 2012. Model SOP for computer forensics. Available at: www.irisinvestigations.com/wp-content/uploads/2016/12/ToolBox/04-ISO QUALITY

SWGDE, 2014a. Digital and multimedia evidence (digital forensics) as a forensic science discipline. Available at: www.slideshare.net/dgsweigert/2014-061120-digital20and20multimedia20evidence20

SWGDE, 2014b. Electric network frequency discussion paper, 14. Available at: www.swgde.org/documents/Current Documents/2014-02-06 SWGDE Electric Network.

SWGDE, 2017. Core competencies for forensic audio. Available at: www.dfir.training/guides/611-swgde-core-competencies-for-forensic-audio/file

SWGDE, 2018a. Best practices for digital evidence collection, 7. Available at: www.swgde.org/documents/Released For Public Comment/SWGDE Collection of Digital.

SWGDE, 2018b. Best practices for digital audio authentication, 27. Available at: www.swgde.org/documents/Released For Public Comment/SWGDE Overview of the.

SWGDE, 2018c. Video and audio redaction guidelines, 16. Available at: www.swgde.org/documents/Released For Public Comment/SWGDE Video and Audio.

Thompson, W.C., 2009. Painting the target around the matching profile: The Texas sharpshooter fallacy in forensic DNA interpretation. Law, Probability and Risk 8, 257–276. Available at: https://doi.org/10.1093/lpr/mgp013

Thompson, W.C. and Schumann, E.L., 1987. Interpretation of statistical evidence in criminal trials. Law and Human Behavior 11, 167–187.

Tversky, A. and Kahneman, D., 1981. The framing of decisions and the psychology of choice. Science 211, 453–458.

Venville, N., 2011. A review of contextual bias in forensic science and its potential legal implications. Australia New Zealand Policing Advisory Agency, Docklands, VIC, Australia.

Chapter 4

The human auditory system

Introduction

The limitations and sensitivities of the human ear are at the core of all audio system designs. This can be highlighted by considering that if they were created for other species, their construction would likely be very different. For example, the upper limit of frequencies perceived by humans extends up to 20 kHz, but cats may find sounds limited in this way a little bass-heavy, as the upper limits of their hearing reportedly extend up to 85 kHz (Heffner and Heffner, 1985). Although the human auditory system is technically continuous, it can easily be modelled by a discrete system due to its various limitations, such as discrete bands and critical masking. As the perception of sound can be perceived in multiple dimensions, each will be considered in isolation to aid in improving their understanding

This chapter will begin by reviewing the human auditory system (or HAS). Going into intricate levels of detail of the ears' anatomy will be of little consequence to the content of this book, so will only be covered briefly. Instead, the focus will be the system's limitations and its perception of sound.

Overview of the human auditory system

The human auditory system begins with the ear, which itself can be separated into three regions based on their specific function. The outer ear consists of the pinna, which is externally visible as the fold of cartilage on the side of the head. The shape of this region is adapted for two functions. First, to provide spatial information caused by the reflections of sound, and second, to direct sound towards the auditory canal, the shape of which causes amplification of sounds in the 3–12 kHz range. Sound then enters the middle ear, which begins with the tympanic membrane (commonly referred to as the eardrum), and is conducted through this region via the malleus, incus and stapes (or hammer, anvil and stirrup). Finally, the inner ear acts as a transducer, ensuring the efficient transfer of sound from the air to internal fluids.

Once sound has travelled through the inner and middle ear, it reaches the oval window of the cochlea, a rolled-up tube containing a membrane of approximately 15,500 hair cells. The cochlea contains the basilar membrane, which has 3,500 sections of four hair cells each, the function of which is equivalent to a 3,500-band frequency analyser. Pressure waves from the stapes cause mechanical resonances along the exponentially distributed hairs of the basilar membrane, which

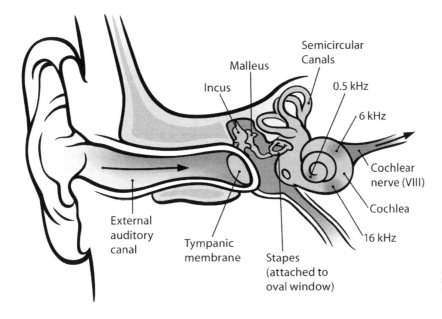

Figure 4.1 Human auditory system
Source: Chittka and Brockmann (2005). Many thanks to the authors for granting permission for the use of this image.

are then transmitted to the brain stem. The area of the membrane nearest the oval window is narrow and contains a dense distribution of hairs, which respond to high frequencies, while the area at the opposite end is wider and contains a sparse distribution, which responds to lower frequencies. As the hairs at the beginning of the membrane are exposed to the most pressure, it is those which are first to break off when damaged (including from the effects of ageing), and as such, sensitivity is lost in the upper-frequency areas of the spectrum first. Hairs cannot grow back, so any loss is permanent (Yamaha Corporation, n.d.). A visual overview of the ear's anatomy is presented in Figure 4.1.

Magnitude perception

Threshold of audibility

The threshold of audibility can be defined as the minimum level at which a sound can be detected in the absence of other external sounds. In order for it to be measured, one of two methods is generally used, either the Minimum Audible Pressure (MAP) or the Minimum Audible Field (MAF) (Churcher and King, 1937; Sivian and White, 1933). MAP measures the sound pressure level at the entrance to the ear canal, or inside the ear using a small microphone. The threshold is then defined as the level at which the listener can perceive the sound (Sivian and White, 1933). MAF uses loudspeakers in an anechoic chamber and measures the level at the position where

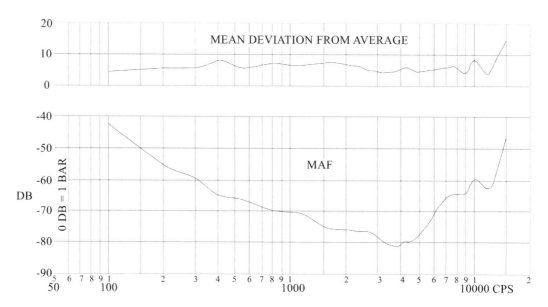

Figure 4.2 MAF values logarithmically averaged for 14 years
Source: Reproduced from Sivian and White (1933).

the head of the subject was located when they detected a sound (Acoustical Society of America, n.d.). Both of these methods use sinusoid signals of a moderate length (over 200 ms) as the sound source. The threshold not only defines the amplitude, but also the frequency range for which sounds can be detected, and it has been shown that there is a high level of correlation between the two (Figure 4.2). This relationship has been published in various studies, including those by the International Organization for Standardization (ISO).

Although not clear from plots (as the length of the sounds used for testing was over 200 ms), the threshold of audibility is affected by sounds of shorter durations in that higher sound intensities are required for detection. The plots do show that we are most sensitive to sound in the 100–5000 Hz region, which also happens to be the region in which the majority of human speech resides.

In terms of magnitude, the lowest level of sound the human auditory system can detect is just 20 micro pascals RMS at 2 kHz, and the term dB SPL actually indicates the pressure of a signal relative to this measurement (Allen and Neely, 1992). The perception of loudness by the human auditory system is logarithmic, and as discussed in Chapter 5, the decibel is used for the measurement of such. The dynamic range between the threshold of audibility and the threshold of pain at the eardrum is approximately 130 dB.

Although understanding the threshold of audibility is helpful as a part of the overall picture of human sound perception, more often than not, our ability to hear a specific sound is limited by noise masking the desired signal. This leads to an increase in the threshold of audibility for that particular sound.

Frequency perception

In terms of frequencies, it is commonly known that there is an upper-frequency limit to our hearing, which can be up to 20,000 Hz for young children, although it decreases with age (Rosen and Howell, 1990), to the point that the majority of people over 30 years of age cannot hear frequencies exceeding 15,000 Hz. Research has found that although there is no lower limit, the way in which we perceive sound changes below 20 Hz. Sine waves from 20 Hz to the upper-frequency limit are detected by the ear as vibrations, but below this, sound may be perceived as a feeling of pressure on the eardrum rather than as a tone (Møller and Pederson, 2004).

Equal loudness contours

In marrying the elements of frequency and amplitude together, the sensitivity of the human ear to sound as a function of frequency was first proposed by Fletcher and Munson (1933) of Bell Laboratories, in a scientific paper which remains highly relevant to this day. Their experiments involved the use of a reference tone of 1,000 Hz, played back via telephone receivers into the ears of the 11 participants in 297 observations. The participants would listen to a pair of tones 1 second apart (the sound being tested followed by a reference signal), twice in a row, pressing a button to indicate their opinion on which was louder. The relative signal levels were then adjusted by an operator and the process repeated until the tones were deemed by the participant to be of equal loudness.

A second experiment was then conducted to transpose the results obtained using the telephone receivers to the real world. A tone was played back in the receivers, then 1½ seconds later played back through speakers at the same voltage. Level adjustments were then made, as in the first experiment, until the loudness of each signal was judged to match. This provided weighting coefficients which could be applied to the data from the first experiment, and a graphical representation of the results, coined Equal Loudness Contours, was created (Figure 4.3).

Following this, further experiments were carried out using the Equal Loudness Contours through the playback of two tones of equal levels at different frequencies. The participants then listened using one ear, followed by both. When listening with one ear, the loudness of the reference tone and that being compared halved, allowing the assumption to be drawn that they must be equally loud when listening with one ear, and as such, result in the same outcome. The findings documented in this paper then led Fletcher to perform further research some seven years later, in which he proposed that sounds are perceived within a series of 'critical bands,' which form the foundation of all present-day perceptual compressed encoders.

Critical bands

While still working at Bell Laboratories, Fletcher published the paper 'Auditory Patterns,' proposing that the ears' sensitivity can be separated into a series of discrete bands, much like a series of bandpass filters (Fletcher, 1940). These bandwidths were termed 'critical bands,' with each being composed of a different shape, bandwidth, centre frequency, and amplitude weighting. Although Fletcher knew the filters were not rectangular, it was convenient for the calculations of masking. The bands, therefore, have a hypothetical flat top, with vertical edges like a perfect filter (which

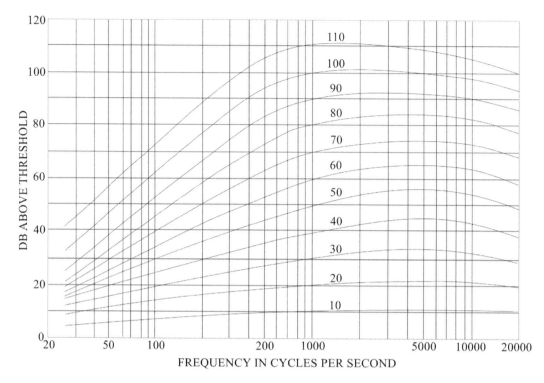

Figure 4.3 Equal loudness contours
Source: Reproduced from Fletcher and Munson (1933).

does not exist, at least in the digital domain). In reality, auditory filters have a rounder top and sloping edges. It was found that regions below approximately 1,000 Hz take on a linear form, but above this become logarithmic, and higher frequencies were suggested to produce greater stimulation of the nerve patches, and consequently a more significant amount of masking. The auditory filters have been shown through multiple research studies to be caused by specific physical locations on the basilar membrane responding to a limited range of frequencies, with each corresponding to a filter with a different centre frequency (Patterson, 1976). The terms frequency selectivity, frequency resolution, and frequency analysis are used interchangeably when referring to critical bands.

Essentially, these findings show that when two signals are produced which occupy the same critical band, the louder tone will 'mask' the quieter tone, causing it to become inaudible. Figure 4.4 provides a basic example of this, although it should be noted that the distribution of the masking frequency is asymmetrical and dependent on the level of the masking signal. At lower levels there is a greater spread of masking towards the lower frequencies than the higher frequencies, and at higher levels the opposite is true. Only at masker levels of 40 dB SPL are masking patterns approximately symmetrical (Zwicker and Jaroszewski, 1982). If the quieter tone is moved into

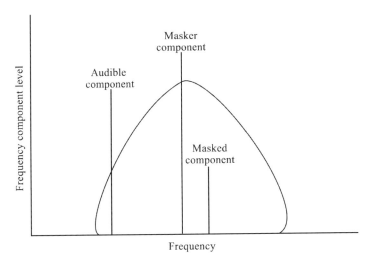

Figure 4.4 Simultaneous masking at approximately 40 dB SPL

an adjacent neighbouring band, it will then become audible. As the perception of sound is highly subjective, it is not possible to propose the exact range of these bands, but attempts have been made to approximate them. The 'Bark' scale is one example. Auditory filters can, therefore, filter the noise to the extent that the only masking is that which occupies the same band as the desired signal. It should be considered that the real world contains complex sounds which will cover multiple bands, of which speech is a prime example.

Masking

Masking is one of the most common causes of poor speech intelligibility and can be either simultaneous or non-simultaneous. It is defined as:

1   The process by which the threshold of audibility for one sound is raised by the presence of another (masking) sound.
2   The amount by which the threshold of audibility of a sound is raised by the presence of another (masking) sound, expressed in dB (American Standards Association, 1960).

Simultaneous masking

The 'cocktail party effect' was coined by Cherry (Cherry, 1953). Imagine standing in the middle of a cocktail party talking to a friend surrounded by 50 people all talking at various volumes and at differing distances from you and your friend. The human auditory system has evolved to allow the brain to focus on the speech of your friend and drown out (or at least ignore) the surrounding babble by using various cues to separate sounds. These include Interaural Level Differences,

Interaural Time Differences, and Interaural Decorrelation (all caused by differences between reflections arriving at each ear in a reverberant environment), the use of lip-reading, different voice characteristics (mean pitches, mean speed, male and female) and differing accents. The human auditory system operates across two dimensions to isolate sound, that of the frequency domain and the time domain. This adaptation works incredibly well in a real-life environment, but once captured as an audio recording, many of these cues are lost, which is a problem for forensic audio recordings containing competing speakers.

Several experiments were performed by Cherry, but the two most applicable to forensic audio are as follows:

- *Two simultaneous messages, mixed.* The first presented two recorded messages of varying similarity mixed together. The task put to the participants was to simply transcribe the speech, replaying the recordings as many times as they wish. The subjects reported great difficulty in accomplishing the task and errors were made, although no long phrases (over two or three words) were wrongly identified.
- *Two simultaneous messages, unmixed.* In this experiment, the messages were isolated and played independently into the right and left ears. It was found that participants had no problem in isolating one message and rejecting the other. In fact, the extent of rejection of the other speech was so strong that, upon being asked to recall anything heard within the message they rejected, they could provide no information about the words, the semantic content or even the language used.

In casework, the majority of recordings containing competing speakers will be mixed, just as they were presented to the listeners of the first experiment, thus making it difficult to reliably determine what was said. In rare cases, and only those captured via specific systems, the voice of each subject may be captured in separate streams (for example, the operator on the left channel and the caller on the right). This is the perfect scenario as simply muting the unwanted channel allows the speech of either person to be replayed with no simultaneous masking whatsoever.

When listening to a signal in noise, the auditory filters focus on the band which the desired sound occupies, reducing the perceived noise surrounding it. In general, for mid-range frequencies, a signal will be detected if its level is no more than 4 dB below the noise present within the same band (Bronkhorst, 2015).

Binaural masking

Binaural masking refers to differences in the signal presented at the left and right ear, although each contains both the target and masking signal. This is opposed to monaural masking in which exact copies of the same signal would be presented to both ears. Research has shown that if the phase of the target signal of a monaural signal is reversed in one ear, the detection threshold reduces by up to 15 dB (Durlach, 1963). One theory proposed by Kock (1950) is the equalisation-and-cancellation model, where the auditory system attempts to eliminate masking through transforming the signal in one ear until the masking components are the same in both ears (equalisation) followed by subtraction of the signal in one ear from the signal in the other (cancellation). As it relates to forensics, this phenomenon demonstrates the advantages of a stereo recording over a single-channel mono recording.

Informational masking

Up to this point, we have only discussed what is defined as energetic masking, a form of masking which physically interferes with the speech signal within the acoustic environment. Informational masking is defined as perceptual masking which occurs due to the limitations of the human auditory system, such as when the masking sound is highly similar to the desired signal, and the properties of such vary unpredictably (Pollack, 1975). In a study by Neff and Green (1987), a 1 kHz sine wave was masked using a number of non-harmonic sine waves of shifting frequencies, and it was found the auditory threshold was much higher than for energetic masking alone. It can be surmised that the desired signal is confused by the similarities between the signals, to the point that the brain cannot perceptually separate them. The reason for such is the inability of the auditory filters to work effectively due to the masking signal transitioning between bands. Two of the biggest causes of this effect on forensic audio recordings are competing speakers and background music. If the music contains a vocal track, effects are further compounded by the transient nature of music and similarities between voices.

In light of this, research performed by Brungart (2001) found that the recognition of speech was highly correlated with the degree of similarity between the target and masker voices, where the greater the level of similarity, the harder it is to distinguish the speech. This is evidenced by the finding that the poorest results were reported when the same voice was used for the target and masker, while the best results were achieved when the sex of the masker was changed.

A second study (Brungart et al., 2001) found that listeners perform better if exposed to the voice of an individual before listening to the combined recording, showing that familiarity with the voice can help. This is something that has not yet been studied for forensic purposes but may have the potential for use by an examiner performing transcriptions of audio containing multiple speakers.

In the real world (although of no use to forensic examiners), we can overcome informational masking through various cues from differing azimuth, distances, and elevation of the desired speaker and the masker. Freyman et al. (1999) showed that when speakers were separated using loudspeakers at 0^0 and 60^0 azimuths in relation to one another, there was a 15–30 per cent improvement in the keywords identified. This is compared to only a 5–10 per cent improvement when the masking speech was changed to noise of the same average spectrum.

Non-simultaneous masking

Masking is not only limited to sounds which occur at the same time as the desired signal. It can also be caused by those occurring slightly before, known as 'forward' or 'pre' masking or slightly after, known as 'backward' or 'post' masking (Figure 4.5) (Deatherage and Evans, 1969).

As the ear has a finite frequency resolution defined by the width of a critical band, it stands to reason that the inverse (or time-domain) resolution must also be finite, something which can be evidenced through Fast Fourier Transform (FFT) theory. The finite temporal resolution explains

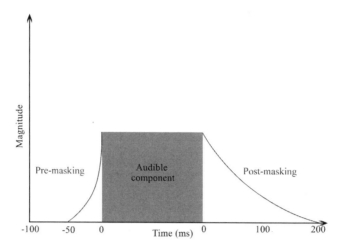

Figure 4.5 Non-simultaneous masking including both pre- and post-masking

why non-simultaneous masking occurs from a theoretical standpoint, but the reason the human auditory system has evolved with what appears to be such an obvious flaw is related to speech intelligibility. Upon processing sounds, our hearing system averages the energy of the signal in frames approximately 30 ms in length. Thus, any reverberation caused by the surrounding environment and occurring within the same frame is added to the signal, resulting in a louder sound. It also reduces the magnitude of subsequent versions of the same signal which occur within approximately 50 ms of the original, known as the Haas or precedence effect (Haas, 1972). If this were not the case, the intelligibility of speech would be much lower as our brain would struggle to distinguish the direct sound from the trailing indirect reflections, resulting in a type of cocktail party effect.

The backward masking phenomenon is poorly understood, despite many research studies having been published. With that being said, it has been shown that the degree of backward masking can be reduced through training, to the extent that those who have practised become unaffected by it.

The following is a summary of the findings from studies into forward masking:

- Forward masking is more significant the closer in time to the masker the maskee signal occurs.
- The rate of recovery is greater for maskers at higher levels, decaying after 100–200 ms.
- Increases to the masker level do not cause linear increases to the amount of masking. For example, increasing the magnitude of the masker by 10 dB may only increase the masking threshold by 3 dB. This differs from simultaneous masking, at least within a band, where increases in the threshold correspond to linear increases to the signal to masker ratio.
- The amount of masking increases with masker duration up to at least 20 ms. Results differ across studies for longer durations.
- Forward masking is influenced by the relationship between the frequencies of the masker and the maskee (as they are in simultaneous masking).

The reason as to why it occurs is still not clear (Moore, 1993).

Spatial perception

One of the primary functions of the auditory system for both humans and animals alike is to provide information to allow us to determine the direction we should focus our visual system's attention. These systems have different mechanisms to enable the localisation of a sound along the planes of distance, azimuth (or angle) and elevation.

The two best-known techniques are the Interaural Intensity Difference (IID), also known as the Interaural Level Difference (ILD), when measured in dB, and the Interaural Time Difference (ITD) (Si, 2016). If we consider a sound source located 10 metres in front and 10 metres to the right of us, the sound will have to travel further to reach the left ear, and so will arrive slightly delayed in comparison to the signal at the right ear. This is the ITD. As it has travelled further, and will also have been subject to refraction and absorption from the head (as it is not a direct path to the far ear from the source), the sound at the left ear will also be of lower intensity. This is the IID/ILD (Figure 4.6). ITD is most useful at low frequencies as the wavelengths of the sound are larger, so differences in phase can be distinguished (Wightman and Kistler, 1992). ILD is relatively independent of frequency (Yost and Dye, 1988).

Another factor is the role of an area of the ear's pinna. These distinctive pieces of cartilage on the sides of the head are crucial to localisation due to the spectral modification they cause, depending on the angle of incidence of the sound relative to the head (Blauert, 1997). Sounds arrive at the eardrum both directly and as delayed reflections from the pinna. Only sounds over 6 kHz are reflected due to their shorter wavelengths and thus interaction with the surface of the pinna. The direct and reflected sounds are then summed in the auditory system and lead to spectral attenuation and boosts. It is these changes that are used by the brain to analyse and locate a signal.

Distance is a little trickier for the HAS to calculate. At face value, the problem is simply this: how do you tell the difference between a sound with a high intensity that is further away, and one with lower intensity but is closer? This is the reason that distance perception is somewhat inaccurate and is generally overestimated for nearby sources and underestimated for distant sources (Zahorik, 2002). The magnitude of sound is useful to some degree if the sound is familiar and there are audible sounds which can be used as a reference, as those at closer distances will have higher ILD than normal (especially for low frequencies). As high frequencies are composed of shorter wavelengths, and thus more prone to reflection, the sound arriving from further away will be attenuated more than lower frequency sounds, but this is only of use if, again, there are other sounds for comparison, or the sound is highly familiar (Coleman, 1962; 1963). When indoors, there are other factors which can be of use, such as the time delay and level differences between the direct and reflected sounds as they provide further information.

Auditory fatigue

Extended periods of exposure to sound and increases to the magnitude of sound can lead to changes in the perception of audio through two distinct processes: adaptation and fatigue.

Adaption of sensory receptors is defined as the decline of the electric responses of a receptor neuron over time, despite the continued presence of an appropriate stimulus of constant strength (Squire et al., 2008). Once the energy used by the receptor exceeds the metabolic energy sustaining it, the process changes to fatigue, and the threshold of audibility increases in a mechanism employed by the human auditory system to protect hearing by reducing its sensitivity to sound

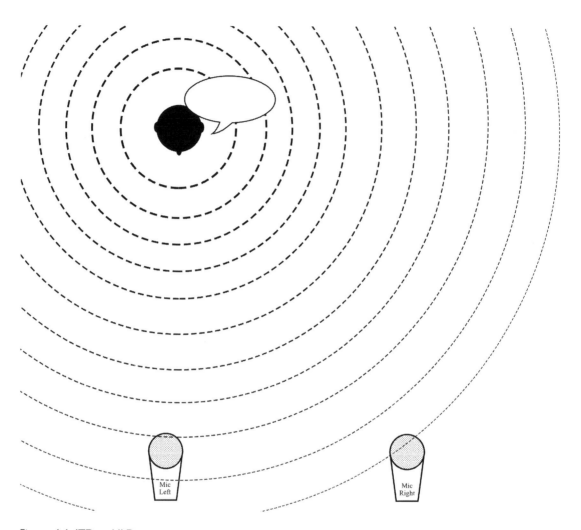

Figure 4.6 ITD and ILD

Note: The thickness of the line represents the magnitude of displacement of the air molecules from normal atmospheric pressure. Note both the difference in the strength and distance travelled of the signals reaching each microphone.

stimulus. This change is known as temporary threshold shift (TTS), 'temporary' as the threshold returns to pre-exposure levels after some time. This is usually within 24 hours, but in some cases, can take up to a week (Kruk and Kin, 2015). Factors which cause adaption and fatigue include the intensity, the duration, and the frequency of the stimulus.

The more serious form of TTS is called permanent threshold shift (PTS), in which the condition becomes permanent, and the auditory threshold never returns to the pre-exposure value.

PTS increases with intensity and duration and is more marked at frequencies around the spectral region of 4 kHz. In the presence of broadband noise, the maximum level of fatigue occurs between 4–6 kHz, and although the effects are small at low levels, they increase rapidly over 90–100 dB SPL (Ryan et al., 2016). Just because a sound appears to return to pre-exposure levels, it does not necessarily mean that permanent damage has not been caused. Generally, sounds above 110–120 dB may produce PTS if over a prolonged duration, which is extremely easy to achieve with modern amplifiers in speakers and headphones and at concerts. A good rule of thumb is to never exceed 50 per cent of the maximum amplitude available within the system used for playback, to mitigate against the risk of PTS, under the assumption that suitable equipment is used, not 1000 w amplifiers. As the average magnitude of a human voice within an acoustic environment is approximately 65 dB, this is the level at which the human auditory system has adapted for maximum intelligibility, and as such, exceeding this provides little benefit (French and Steinberg, 1947). With that in mind, a more precise approach would be to measure the SPL at the ear to understand how loud 65 dB is, and to never exceed around 70 dB.

Summary

The human ear is where all audio forensics enhancement work is primarily destined, whether it be the ear of the judge, jurors, or legal representatives. Only by understanding the auditory system and its limitations can we ensure approaches to various tasks are optimised for the end user. This chapter also serves to highlight the potential dangers of sound, and mitigating against any hearing damage should be high on the list of an audio examiner's priorities to ensure both a long career and the ability to extract as much information as possible from every recording.

References

Acoustical Society of America, n.d. 4.43 minimum audible field. Available at: https://asastandards.org/Terms/minimum-audible-field/

Allen, J.B. and Neely, S.T., 1992. Micromechanical models of the cochlea. Physics Today 45, 40–47.

American Standards Association, 1960. Masking.

Blauert, J., 1997. Spatial hearing. MIT Press, Cambridge, MA.

Bronkhorst, A.W., 2015. The cocktail-party problem revisited: Early processing and selection of multi-talker speech. Attention, Perception & Psychophysics 77, 1465–1487.

Brungart, D.S., 2001. Informational and energetic masking effects in the perception of two simultaneous talkers. The Journal of the Acoustical Society of America 109, 1101–1109.

Brungart, D.S., Simpson, B.D., Ericson, M.A., and Scott, K.R., 2001. Informational and energetic masking effects in the perception of multiple simultaneous talkers. The Journal of the Acoustical Society of America 110, 2527–2538.

Cherry, C.E., 1953. Some experiments on the recognition of speech, with one and with two ears. The Journal of the Acoustical Society of America 25, 975.

Churcher, B.G. and King, A.J., 1937. The performance of noise meters in terms of the primary standard. Journal of the Institute of Electrical Engineers 81, 59–90.

Coleman, P.D., 1962. Failure to localize the source distance of an unfamiliar sound. The Journal of the Acoustical Society of America 34, 345–346.

Coleman, P.D., 1963. An analysis of cues to auditory depth perception in free space. Psychological Bulletin 60, 302–315.

Deatherage, B.H. and Evans, T.R., 1969. Binaural masking: Backward, forward, and simultaneous effects. The Journal of the Acoustical Society of America 46, 362–371.

Durlach, N.I., 1963. Equalization and cancellation theory of binaural masking-level differences. The Journal of the Acoustical Society of America 35, 1206–1218.

Fletcher, H., 1940. Auditory patterns. Review of Modern Physics 12, 47–65.

Fletcher, H. and Munson, W.A., 1933. Loudness, its definition, measurement and calculation. The Journal of the Acoustical Society of America 82–108.

French, N.R. and Steinberg, J.C., 1947. Factors governing the intelligibility of speech sounds. The Journal of the Acoustical Society of America 19.

Freyman, R.L., Helfer, K.S., McCall, D.D., and Clifton, R.K., 1999. The role of perceived spatial separation in the unmasking of speech. The Journal of the Acoustical Society of America 106, 3578–3588.

Haas, H., 1972. The influence of a single echo on the audibility of speech. Journal of the Audio Engineering Society 20, 146–159.

Heffner, R.S. and Heffner, H.E., 1985. Hearing range of the domestic cat. Hearing Research 19, 85–88.

Kock, W.E., 1950. Binaural localization and masking. The Journal of the Acoustical Society of America 22, 801–804.

Kruk, B. and Kin, M.J., 2015. Perception of timbre changes vs. temporary threshold shift. Paper presented at the 138th AES Convention, Audio Engineering Society, Warsaw, Poland.

Møller, H. and Pederson, C., 2004. Hearing at low and infrasonic frequencies. Noise and Health 6, 37–57.

Moore, B.C.J., 1993. Characterization of simultaneous, forward and backward masking. Paper presented at 12th International Conference: The Perception of Reproduced Sound, Audio Engineering Society, Copenhagen, Denmark.

Neff, D.L. and Green, D.M., 1987. Masking produced by spectral uncertainty with multicomponent maskers. Perception & Psychophysics 41, 409–415.

Patterson, R.D., 1976. Auditory filter shapes derived with noise stimuli. The Journal of the Acoustical Society of America 59, 640–654.

Pollack, I., 1975. Auditory informational masking. The Journal of the Acoustical Society of America 57.

Rosen, S. and Howell, P., 1990. Signals and systems for speech and hearing. Brill, Dordrecht.

Ryan, A.F., Kujawa, S.G., Hammill, T., Le Prell, C., and Kil, J., 2016. Temporary and permanent noise-induced threshold shifts: A review of basic and clinical observations. Otology & Neurotology 37, 271–275.

Si, W., 2016. Relative contribution of interaural time and level differences to selectivity for sound localization, Paper presented at the 140th AES Convention, Audio Engineering Society, Paris, France.

Sivian, L.J. and White, S.D. 1933. On minimum audible sound fields. The Journal of the Acoustical Society of America 4, 288.

Squire, L.R., Bloom, F.E., Spitzer, N.C., et al. (eds) 2008. Adaption of sensory receptors. In Encyclopedia of neuroscience. Academic Press, New York.

Wightman, F.L. and Kistler, D.J., 1992. The dominant role of low-frequency interaural time differences in sound localization. The Journal of the Acoustical Society of America 91, 1648–1661.

Yamaha Corporation, n.d. Human auditory system. Available at: uk.yamaha.com/en/products/contents/pro-audio/docs/audio_quality/04_audio.

Yost, W.A. and Dye, R.H., 1988. Discrimination of interaural differences of level as a function of frequency. The Journal of the Acoustical Society of America 83, 1846–1851.

Zahorik, P., 2002. Auditory display of sound source distance. Paper presented at the 2002 International Conference on Auditory Display, Kyoto, Japan.

Zwicker, E. and Jaroszewski, A., 1982. Inverse frequency dependence of simultaneous tone-on-tone masking patterns at low levels. The Journal of the Acoustical Society of America 71, 1508–1512.

Chapter 5

Key audio principles

Introduction

Although it is crucial that all of the fundamental concepts relating to digital audio are understood, to avoid redundancy, only those which most commonly apply to audio forensics will be discussed here. With that in mind, there will undoubtedly be areas in the science of audio which could be, and possibly are, exploited for use in audio forensics that will not be covered. The reader is, therefore encouraged to seek out further sources to assist in the advancement of knowledge beyond the fundamentals.

Many of the concepts are available in numerous other books, so only an overview will be given of each, in order not to stray too far from the focus of the book.

Forensic audio versus studio audio

Although from the outside looking in, audio forensics laboratories and music recording studios may appear to be very similar, both the products and principles used could not be further apart. In addition, there is only a small region of overlap in the type of work performed between the two. For example, recording studios do not authenticate recordings, and forensic laboratories do not record audio, but both perform enhancements, albeit for different purposes. With that being said, the key concepts are the same and are all highly applicable to the processes undertaken.

The type of work performed within each scenario is dictated by a combination of the data on which the processes are applied and the end user requirements. Sound engineers work with audio that has been recorded in an environment optimised for the characteristics of the specific sound being captured, such as reverberant studio live rooms and near-anechoic vocal booths. Forensic examiners are provided with recordings which could not be further from the hi-fidelity audio processed by a recording studio. These types of recordings are captured using less than ideal equipment (think phones with electret microphones and telephone transmissions), often by untrained individuals, and in some of the most unsuitable acoustic conditions possible. The purpose of the role is also completely different. Whereas the studio engineer is fixated on providing a sound that will ultimately entertain the end listener, a forensic audio examiner is focused on extracting as much information as possible from a recording, the results of which are provided to those within the justice system, whether it be an enhanced sentence or the presence of an edit point.

To provide the most explicit demonstration of this disparity, each element of variation will be addressed separately (Zjalic, 2018).

Capture environment

Studio audio

A large portion of studio work pertains to the capture, while the mixing occupies the remainder. The recording element is done within a live room or vocal booth, both of which are highly controlled environments designed for the optimal capture of sound. There are, therefore, no superfluous sounds to detract from the desired signal, and the signal to noise (SNR) ratio is exceptionally high. An audible HVAC (heating, ventilation, air-conditioning) system would be high on the list of a studio engineers' worst nightmares, and if this were captured on recordings, the room would be considered unusable. For the forensic examiner, if the presence of a HVAC system on a recording is the only issue, it would be a dream come true, compared to the other quality issues faced. Studio recording is focused on getting the best sound during capture to minimise the amount of processing required during the mixing stage. Although some engineers may have the attitude to fix it in the mix, this is not an approach taken by those who are experienced or work in high-end studios.

Forensic audio

Unfortunately, forensic examiners have no influence over the capture, so the recordings are always less than ideal as they are captured in uncontrolled environments, which contain extraneous content to that which is desired. Some common examples include recordings made in venues containing crowds of people talking and loud background music, recordings made by emergency call centre systems of some form of violent altercation (often containing shouting, sirens, screaming, and subsequently clipping) and recordings captured two rooms away from the related event. All of these factors result in low signal-to-noise ratios, which can extend from negative (indicating the noise is of a higher amplitude than the desired signal) up to approximately 40 dB.

Equipment

Studio audio

As recording studios are designed to capture multiple simultaneous inputs, they always contain a patch bay, a mixing console, and a multitrack recorder. Further to this, there are potentially racks of processing units, such as compressors, reverb units, and equalisers, as studio recordings are often captured in a dry environment and have processing applied post-capture to bring them to life. This allows an increased level of control over the sounds, as it is much easier to start with a dry sound and add a process (such as reverb) than start with a reverberant recording and attempt to remove it. Mixing takes place in an acoustically treated control room to reduce reflections which can muddy the sound during playback. Both the live rooms and control rooms are always built away from environmental noise sources, such as train tracks and roads, which would not only be captured while recording, but also add external noise to the room during playback and mixing.

Forensic audio

One of the only common requirements of both a studio control room and a forensic laboratory is the playback of audio. Forensic labs generally do not need equipment to capture audio, but as they may need to convert it from analogue formats such as cassette tapes, there will usually be a tape player connected to the input of an audio interface. Although some forensic processing equipment can come in the form of rack-mounted hardware, it is not uncommon to see this in a laboratory, but the majority is now software-based. There is no requirement for monitor speakers of any kind as headphones are the primary listening method, but speakers may be used to allow multiple examiners to listen to a recording to facilitate a discussion relating to the audio, before or after the work has been performed. The reason headphones are the primary method of listening is their ability to prevent extraneous external noise being added to already poor-quality recordings. It is common to switch between headphones to prevent dips which may occur due to the combination of the outer ear and headphone. This also reduces the 'anchor effect,' a term for the influence of past experiences (such as the frequency response of single set of headphones) that are used consciously or subconsciously as reference points for current judgements (Letowski, 1992). As with a recording studio, the laboratory is usually away from potential sources of external noise. As the audio is submitted on external devices, or provided in the form of external media, there will be the necessary equipment to facilitate the transfer of the audio, including data cables, USB ports, memory card readers, CD/DVD readers, and most importantly, write-blockers. These exist in both software and hardware variations to ensure recordings are imaged to the workstation in a forensically sound manner and prevent changes being made to the original version on the device or the external memory.

Microphones

Studio audio

Unsurprisingly the function of recording studios is in the name. Recording. Their primary role is the capture of audio, so microphones are of the highest quality, and can be carefully selected based on the type of sound being captured, and the characteristics of the sound desired. It is also common to use arrays of microphones in various stereo formations to ensure the entire soundscape is captured to the optimal level. Microphones are also uncovered and positioned with consideration for the sound source.

Forensic audio

In general, the microphones used in the digital audio forensics field have limitations imposed by requirements relating to size, cost, and convenience. If an individual is attempting to capture an undercover recording, the last thing they want is a recording device with the dimensions of a house brick attached to their waist, as depicted in films from the 1970s, so the smaller, the better. Technology has brought about improvements in sound quality while simultaneously reducing the size of devices to the point that the dimensions of the most miniature of microphones in use today are smaller than coins. This does limit the input to some extent, although it is possible to have small diaphragm microphones with an excellent frequency response. In terms of cost, the environment is often uncontrolled, and in many cases, the device may only be purchased for the task at hand, so it is difficult to justify spending large sums of money on a capture device. The vast majority of recordings do not seek to capture a crime intentionally, so in reality, devices are not optimised for the sound they do capture. For the lower-priced models, this dictates the quality of the components, as opposed to expensive studio microphones which are used daily in a safe environment and are not at great risk of damage. More pertinent than ever is the matter of convenience. The most widely used recording devices are those which are a by-product of a device everybody carries with them at all times: a mobile phone. It makes sense that these devices are the go-to recording device for most people. As their primary use is making and receiving telephone calls, their ability to capture high-quality audio is somewhat limited. It is often the case that people do not want others to know they are recording, particularly in cases covert in nature, so the microphone diaphragm is covered. This can lead to lower magnitude levels and a sound with a muffled character due to the absorption of various frequency components by the item covering the microphone. It can also lead to noise caused by clothing hitting against the diaphragm of the microphone, which invariably causes simultaneous masking of the desired signal.

Operators

Studio audio

Sound engineers are academically explicitly trained in the areas of recording, mixing, and mastering of audio or have accumulated experience shadowing other engineers in the studio, or a mixture of both. This, in turn, leads to individuals who know precisely which microphones to use, where to place them, and the specifications to use for a specific type of recording to ensure the optimal capture of audio.

Forensic audio

Operators in the forensic audio field are, at best, individuals who have some form of understanding of how to capture sound, such as undercover agents working for government security services. At worst, and as is more often the case, the people who capture audio recordings are those with no knowledge of audio concepts whatsoever. This results in recordings captured using low-quality devices, in acoustically poor locations, in a compressed format. A real casework example is the use of a mobile phone capturing audio in a low bit rate MP3 format placed under a bed in a house, with the gain setting at a maximum in an attempt to capture a spouse's every action while they were out. Recordings captured by an audio device in a trouser pocket or an inside jacket pocket are also a prevalent scenario, resulting in a broadband rustling sound caused by even the slightest movement, due to the noise generated by the clothing rubbing against the microphone.

Specifications

Studio audio

Studio recordings are generally captured in an uncompressed PCM format at a minimum sample rate of 88.2 kHz and a 32-bit floating-point bit depth, to ensure the highest quality possible. This can result in large file sizes, but as there is no need for the device to be portable, this is not an issue as the memory capacity of a computer is virtually limitless.

Although the recording will most likely be transcoded to a compressed format later down the line to aid in streaming and downloading from the internet, the material that sound engineers work with is of optimal quality and is never degraded in any manner.

Forensic audio

The majority of forensic audio recordings are of a lossy compressed, low specification format due to a culmination of two of the factors previously discussed. First, the portable nature of most devices means storage is at a premium, so compression aids in allowing a considerable amount of data to be stored through sacrifices to quality. Second, those making recordings generally do not have an understanding of the audio formats and specifications which would allow them to make an informed decision on the best settings to use when presented with several options. I would assume that as MP3 is the most popular format in the world, if given a choice, most people will go with MP3, which they are most familiar with, rather than one which has the highest specifications.

PCM recordings from portable recording devices are often used in situations where the recording was premeditated, rather than a spur-of-the-moment capture as an incident unfolded. The most likely reason for this is the use of devices specifically designed for covert audio capture which use uncompressed formats. They are also common for emergency service call captures as there is no need for the data to be stored in a compressed format to save space.

Sample rates of forensic audio generally extend from 8 to 48 kHz. Still, in reality, this is of little consequence in the majority of cases, especially considering that the bulk of speech data exists below 4 kHz, thus requiring a minimum 8 kHz sample rate.

Devices

Studio audio

Studio recordings use equipment which remains within the same environment, which is usually the same space where the recording was captured. The system is extremely stable, thanks to its static nature, and as such, issues relating to signal dropouts and electrical interference are highly unlikely.

Forensic audio

As it applies to forensic audio, the location in which the recording is captured can be stationary or in transit, dependent on the device used. When moving, the chance of damage to the equipment or electrical interference becomes increasingly likely when compared to a static system.

End user

Studio audio

The end user of the final product created within a studio recording is the general public. A studio audio engineer aims to provide a result that will entertain, whether it is a song played on the radio

Table 5.1 Studio audio and forensic audio characteristics compared

Characteristic	Studio audio	Forensic audio
Frequency response	20 Hz to 20 kHz, or better	20 Hz to 20 kHz
Signal-to-noise ratio	90 dB +	Negative to 40 dB
Distortion	Negligible	Negligible to severe
Equipment operator	Trained technician	Investigator, untrained member of the public
Microphone	Small to large professional	Miniature
Medium recorder	Professional digital	Inexpensive to professional digital
Medium quality	Best	Standard
Medium type	Removable, flash memory or hard drive	Removable, flash memory, or hard drive
File compression	Usually none	None to highly compressed
Reverberation	Usually dampened	Often high
Mic to speaker distance	Close	Varies
Microphone location	Open	Concealed
Transmission system	Usually none	Telephone, low power RF or none

or a voice-over for a film. There is, therefore, no requirement for the result to be a true representation of the sound captured, or for the process to be repeatable. The sound will be manipulated to meet the desires of the listeners, artist, and engineers.

Forensic audio

The end user of work performed by a forensic audio examiner is the justice system, whether in the criminal, civil, commercial, or private domain. It is, therefore, of the utmost importance that any generated exhibits are representative of the content of the audio recording. This can be ensured by maintaining the chain of custody, validating methods, and taking contemporaneous working notes. The work is independent and is not influenced in any way by the instructing party, examiner, or potential end users.

Table 5.1 shows a comparison of studio audio against that of forensic audio to summarise the differences. The original version of this was published in (Koenig et al., 2007), but some updates have been made due to the continually evolving nature of technology.

Although dealing with the same medium, it should be evident that there are significant differences in several key areas between the job roles of studio audio engineers and forensic audio examiners. The overlap between the two disciplines is therefore much less pronounced than people may think, to the point that there are probably more differences than similarities.

Capture

The capture and playback of sound can be considered to be a series of energy conversions performed by several transducers at differing points in the process. The stage at which the audio is stored is analogous to a mirror in that the playback process is simply the capture process in reverse (Figure 5.1).

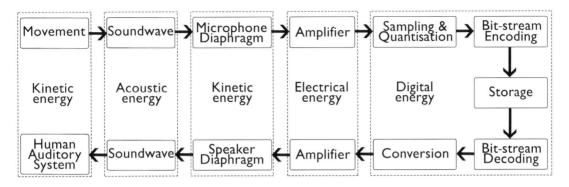

Figure 5.1 Audio capture and playback process with the associated changes to the type of energy used to transmit the information at each stage

For audio to be captured, there must first be an acoustic vibration. This may be the sound of the gunpowder escaping the barrel of a pistol or the vocal cords of an individual making an emergency service call to report the sound of the firearm on a street. These vibrations disturb the surrounding air molecules, commonly analogised in the world of sound as the distribution of waves when a stone is dropped into a pond. These waves disperse spherically in all directions from the source of vibration, and both the size of the waves and period between each are dependent on the degree of displacement caused by the initial vibration (Figure 5.2) (Rossing et al., 2013).

The repetitive nature of these waves lends them to be defined by their 'frequency,' or the number of times that they repeat over a period of time. Although sounds do not exist as isolated sine waves, for now, we will assume they do to simplify some of the concepts. For the frequency to be measured, the 'period' of time it takes for complete one cycle of a wave needs to be defined. A standard measurement of time is also required, which, in the world of audio, is always 1 second. Once the time and period have been established, the frequency can be easily calculated. For example, imagine an elapsed period of 0.5 seconds between two identical regions within a wave passing a physical location within a room.

Frequency = Time / Period

Frequency = 1/0.5

Frequency = 2 samples per second (Hertz)

The more repetitions of a signal over time, the higher the frequency of the signal.

It is the sound waves hitting the diaphragm of the microphone (a thin sheet composed of an exceptionally lightweight material) that cause it to move. The degree of movement is then converted to an electrical representation of the amount of displacement, the method of which is defined by the type of microphone.

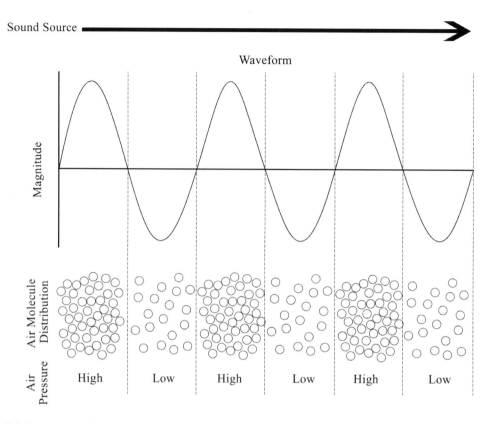

Figure 5.2 Propagation of a sound source demonstrating the waveform representation

The ADC or Analogue to Digital converter can be thought of as the system which takes an analogue input (electrical energy at the microphone stage) and converts the information to the digital domain. The system itself has a number of elements, all of which occur in the opposite order when the audio is replayed and converted back to the analogue domain by way of the DAC, or Digital to Analogue converter.

Acoustic sound exists in two dimensions, time and space, or more precisely, a period of each wavelength (time) and the displacement of each wavelength from the atmospheric pressure (space). It, therefore, makes sense for the ADC to represent these dimensions when capturing the audio as the sample rate (time) and bit-depth (space) (Huber and Runstein, 2005).

Sample frequency

Sample frequency refers to the number of samples captured per period (most commonly defined in Hertz in digital audio). Whereas the real-world movement of waves is a continuous phenomenon (imagine an expensive Swiss watch with a smooth, continuous rotation), sampling is discrete by nature and can be analogised as a digital watch, where there are only a specific number of increments to define the time represented by the expensive watch. For the digital clock to show the time, samples of the continuous-time representation displayed by the expensive watch must be taken periodically, thus mapping the time to a discrete range of available values.

As digital audio takes samples of continuous, analogue sound at a rate of 8,000 times per second or more, there are no issues relating to humans perceiving the discrete nature of digital sampling. When compared to films captured at 25 frames per second (a sample rate of 25 Hertz), audio has extremely high rates.

So, what happens when a signal with a frequency above that of the sample rate is sampled, and only one sample is taken between the start and end of a periodic wave is taken? How does the system know the period if it cannot identify two matching points on the waveform? The answer is that it doesn't.

It was this problem that was addressed by Nyquist (1928), who discovered that for a signal to be represented accurately by a system, it must be sampled at least twice per wavelength. When translated to sample rates, this means the highest frequency within a system cannot exceed half the sample rate. The frequency at 50 per cent of the sample rate is therefore known as the Nyquist frequency, and the concept is called the Nyquist Theorem. Recordings in which the frequency components do not exceed this point are said to 'meet the Nyquist criteria.'

If signals which do not meet the criteria are sampled, the result is a phenomenon called 'aliasing,' in which signals exceeding half the sample rate are misrepresented as 'alias signals.' In order to prevent this, an 'anti-aliasing filter' is applied to the signal before the sampling stage (Figure 5.3). This is basically a low pass filter with a cut-off around the Nyquist frequency (Audio Engineering Society, 2018).

Bringing together the above in a practical sense, this is one of the reasons for the standard 44.1 kHz sample rate. As the upper-frequency limit of the human auditory system is approximately 20 kHz, a 40 kHz sample rate is required to represent all of the signals in this range accurately. The extra bandwidth is applied to allow a smooth roll-off to minimise the effects of ripple (classed as undesirable deviations from the response of an ideal filter) and the transition of the low pass filter. The reason for the odd decimal point is to meet the requirements of video systems, as covered earlier in the book. The Nyquist Theorem is also behind the selection of the 8 kHz sampling rate for telephony applications, as the majority of speech information exists below 4 kHz. The sampling rate used within telephony also has bandwidth restrictions, as sampling at a higher frequency would increase the bandwidth, and thus the data rate required for transmission. The 8 kHz sample rate is, therefore considered an effective solution.

Quantisation

Quantisation is the process of defining data from a broad range as values within a smaller range, analogous to the previously mentioned representation of a smooth Swiss clock as a digital clock. It is the broad range of values used by the Swiss clock that results in the fluid motion as the incremental movements of the hands are so small the human eye cannot perceive them. As digital data is finite and real-world acoustic sound is infinite, quantisation is an unavoidable step during the capture of digital audio.

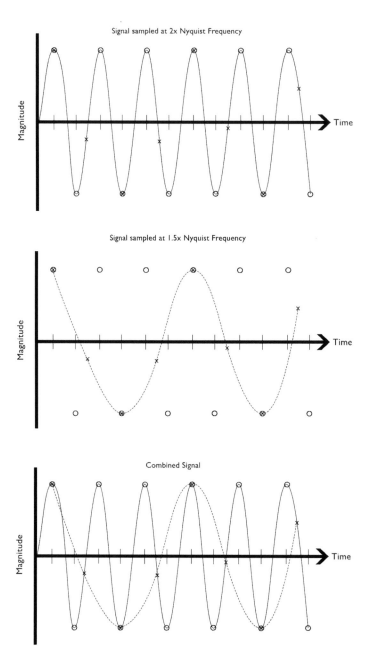

Figure 5.3 Sampling theorem visualised. Upper: Signal sampled at 2x Nyquist frequency, Middle: Signal sampled at 1.5x Nyquist frequency (and thus aliasing and not a true representation of the original signal), Bottom: Signal containing both accurate and aliasing components

These discrete values available within the digital system are defined by the bit depth, which represents the magnitude of a continuous electrical signal (which was previously converted from the displacement of the microphone diaphragm) in the form of a log two number. As the number of levels can become exponentially large, the concept is best explained at low bit depths, as once the theory is understood, it is easily scalable. It is possible to have both linear and non-linear quantisation, but so as not to confuse things it will first be assumed that all quantisation is linear.

The most straightforward starting point is to imagine a system with a bit depth of just a single bit. As 1^2 is 2, all data values will be quantised to one of two binary states based on the value that the input signal is closer to. Next imagine a system with a bit depth of 2 bits, so there are now four values available to represent the signal level (2^2), all of which are linearly spaced, just as the sample rate is. For simplicity, imagine the levels represent -1, -0.5, 0.5, and 1. Zero is also a level (in which there is no signal). As these are the only levels available, the signal entering the system must be rounded to the nearest value. For example, 0.45 would be represented as 0.5, and -0.8 as -1 (Figure 5.4). The disparity between the input value and that to which the value is quantised is known as the quantisation error and is defined as the furthest possible point at which a quantised value is from its input value, or +/-0.5 the quantisation step size. The numerous distortion elements produced by the quantiser become continuous across the bandwidth of the signal and are perceived as low-level white noise (Maher, 1992), known as quantisation noise. When comparisons are made against the original signal before quantisation, the data added to the original signal in the form of quantisation error is visible.

There are several factors which affect the noticeable presence of the quantisation error on the signal. First, the magnitude of the quantisation error is directly correlated with the error rate, and, as such, a higher number of quantisation steps can reduce the signal-to-quantisation noise ratio, and subsequently reduce its impact on the signal. A second factor is the degree of variation within the signal. A third is the magnitude of the signal. Although there may be a large number of steps available, if a signal only occupies a small range of these, for example, when a low magnitude signal is captured, the quantisation error becomes non-random. It is then easily discernible to the human ear as harmonics, sub-harmonics, aliased harmonics, and intermodulation (Katz, 2002). It is the correlation of the quantisation error with the signal, which means it is classified as a distortion rather than noise. As the quantisation process occurs after the anti-aliasing filter, any harmonics which exist above the Nyquist frequency will cause aliasing and be audible as spurious tones analogous to bird song. If the sample rate is a multiple of a single input frequency, the result is harmonic distortion. If there is more than a single frequency present at the input, intermodulation occurs, causing an effect known as granulation noise, perceived as a gritty, metallic timbre (Maher, 1992).

To limit the perception of quantisation noise by the human auditory system, a process known as dithering is applied to the signal during encoding. The theory behind dither is as follows: if the values of a signal are continually moving within a broad range, the quantisation error will be random rather than a consistent tone. Humans are highly efficient at discerning tones, so the more random the signal, the less likely it is to be noticed. To increase the variability of the quantisation noise, a second, pseudo-random noise signal can be added to the input of the quantiser and subtracted at the quantiser output. The process works as it breaks up the signal-dependent patterns without increasing the variance of the error (Jayant and Rabiner, 1972).

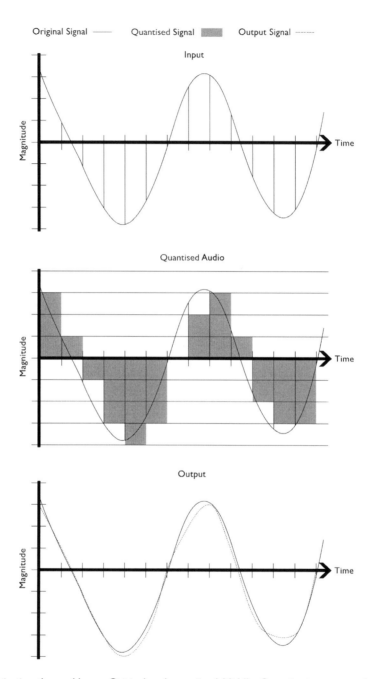

Figure 5.4 Quantisation theory. Upper: Original analogue signal, Middle: Quantisation process, Lower: Differences between the analogue and digital signal

For example, consider the following series of consecutive samples values:

0.6	1.3	1.2	0.9	0.6

When quantised, this would result in:

1.0	1.0	1.0	1.0	1.0

which is highly correlated and thus, audible to the human ear. When a second signal is created, as follows:

−0.2	0.3	0.1	0.8	−0.3

And subsequently added to the signal before quantisation takes place, the result is:

0.4	1.6	1.3	1.7	0.3

which, when quantised leaves:

0.0	2.0	1.0	2.0	0.0

Thus, it is no longer sequential, and the perception of the consecutive matching samples by the human auditory system is reduced.

A signal level above that of the highest level available (e.g. 1.3), would be reduced (or 'clipped') back down to the maximum value available within the system to represent the value, which for the previous example would be 1. If there is a consecutive series of samples which exceed the maximum quantisation level available, the human ear will identify a distinct pattern of successive values in the form of audible distortion.

All quantisation steps mentioned so far in this chapter relate to uniform quantisation, but for perceptually encoded audio, non-linear quantisation is used to enable the most efficient use of bits for storage and to aid in preventing quantisation noise becoming audible. This involves the quantisation of the outputs of a filter bank (essentially a series of bandpass filters) by a finite number of levels while minimising the quantisation noise. A uniform quantiser will use the same step between each level throughout the range, whereas, for a non-uniform quantiser, the quantisation noise is varied in relation to the input value. This is more representative of the logarithmic nature of the human auditory system (Painter and Spanias, 2000). Estimating the quantisation noise of a uniform quantiser is relatively simple as the range of available quantisation steps determines the quantisation error. For non-uniform quantisation, the input values have more of an impact on the quantisation noise than the range. MPEG 1 layer 3 (MP3) and MPEG-2 AAC, and thus MPEG 4 AAC all use scale factors to determine the acceptable amount of quantisation noise based on coefficients (Sreenivas and Dietz, 1998) and thus inform the non-linear quantisation. All quantised audio data can be represented using various types of measurements, such as decibels, power spectral density, or normalised value.

Sampling and quantisation are two key concepts at the heart of digital audio recording in that a sample of the acoustic waveform is taken, and the sample then quantised to the nearest level. Although one might think that the result would be coarse due to the discrete nature of the sampling and quantisation processes, it should be considered that standard CD-quality audio is sampled 44100 times per second with 65536 quantisation levels available to represent the displacement of the acoustic waveform.

Channel configuration

The sampled and quantised audio data is represented as a sequence of values, where each value pertains to a single sample of the acoustic waveform. A single sequence can also be considered as a single channel which can be replayed in isolation, or in conjunction with other channels and stored within a file. Although files do exist which replay larger numbers of channels simultaneously (e.g. 5.1 surround sound consists of six channels), it would be rare to encounter a recording with over two channels in forensic audio. The configurations of these channels most commonly encountered are as follows:

- *Monaural*: This is a single channel of audio data. It may have been captured using a single microphone or captured using two microphones and subsequently converted to a single channel in post-processing. Upon playback, a copy of the single-channel will be sent to both earpieces if using headphones, or both speakers if using a stereo pair of such.
- *Stereo*: This has its roots in experiments conducted in the early 1930s by Alan Blumlein, an E.M.I. researcher (BBC Radio 4, 2008), and pertains to two or more channels of audio data (which are not necessarily different). True stereo contains different signals within each channel, each relating to a separate microphone, which are sent to separate ears during headphone playback, recreating the spatial information of the capture environment. The differences between the channels are caused by Interaural Level Differences and Interaural Time Differences. When two microphones are placed at a distance from one another, there will be small differences between the sound entering each, in terms of both level and delay. Imagine a voice to the right of a microphone array. The sound arriving at the far microphone has further to travel, so there will be a delay (ITD) in the time it takes for the wave to reach the microphone in comparison to the closer microphone. More energy will have also decimated in doing so, causing the difference in level (ILD). This information can be exploited for several areas of forensics, including analysis of acoustic environments and sound source localisation. For example, if a pair of microphones are static, analysis of the differences between the channels can provide information as to the direction from which the sound arrived.
- *Dual mono* pertains to a single channel which has been duplicated, rather than two channels which contain different information. The most likely scenario under which this occurs is the conversion of a single channel to two channels post-capture. This technique has its uses in entertainment, such as for panning sounds to create a perception of an acoustic space, but the level of redundancy would be extremely high if a device with a single microphone was to store the data as two identical channels. It is therefore only really found within audio which has undergone conversion, either intentionally so panning and space can be added, or non-intentionally when it is required to meet the criteria of a particular file specification when converted. Playback is essentially the same as listening to a monaural recording where the

corresponding data is sent to both headphone earpieces or speakers if no processing of each channel for creative reasons has been applied.

Bit rate calculation

Bits are the binary digits used to store digital information, and 'rate' is the period of time over which an event occurs. Bit rates can, therefore, be defined as the amount of data stored or transmitted per second of time. It is calculated from the specifications and has an impact on the amount of data required to store signal information, namely, the familiar concepts of sample rate, bit depth, and channel configuration. The formula is as follows:

bit rate (bits per second) = sample rate (in hertz) x bit depth x number of channels

Once calculated, the value can then be converted to kbps. Traditionally, as computers use log2 calculations, computer scientists equated 1 kB to being equal to 1,024 bytes (2^{10}). This has since undergone a transition to the decimal representation, attributed to the widespread use of computers by non-scientists. In 1998, NIST (the National Institute of Standards and Time) and the IEC (International Electrotechnical Commission) altered the definition of a kB to be equal to 1,000 bytes, thus reducing any potential confusion between figures stated by different companies and individuals (Barrow, 1997). With that being said, computer operating systems still report using the binary representation.

By way of an example, the bit rate calculation for an audio recording consisting of two channels with a sample rate of 44100 Hz and a bit depth of 16 is as follows:

bit rate = sample rate x bit depth x number of channels

bit rate = 44100 x 16 x 2

bit rate = 1,411,200 bits per second (bps)

bit rate = 1,411,200/1024

bit rate = 1,378.125 kilobits per second (kbps)

bit rate = 1,378.125/1024

bit rate = 1.35 megabits per second (mbps)

To obtain the size of the audio stream:

audio stream size = bit rate x duration in seconds

The result can then be converted to bytes using the following (as there are eight bits in a byte):

audio stream size in bytes = audio stream size in bits/8

The encoding bit rate is that at which an audio recording was encoded per second, calculated as follows:

encoding bit rate (bits per second) = (file size (bytes)/duration (s)) x 8 (multiplied by 8 to convert bytes back to bits)

Therefore, for a recording of approximately 9 seconds with a file size of 143180 bytes:

encoding bit rate = (143180/9) * 8

encoding bit rate= 127271 bps

encoding bit rate = 127.3 kbps.

If the true bit rate of the data pertaining to the encoded audio is required (excluding the container and other data streams), only the size of the audio payload should be calculated. The bit rate may also be represented in bps (bits per second), kbps (kilobits per second) or Mbps (Megabits per second), dependent on the magnitude of the bit-rate. When calculated by software, the encoding bit rate will generally be in the form of a constant bit rate (CBR), in which the number of bits used per sample is constant across the recording. A variable bit rate (VBR) changes based on the amount of information within each sample. It is therefore provided as the average across all samples, although information relating to the minimum and maximum bit rates may also be given.

In general, the higher the bit rate, the higher the quality of the audio due to the availability of more data to represent the captured sound and thus less data pertaining to the raw PCM input is removed during the compression/encoding stage. For example, MP3 audio encoders offer a range of bit rates for the storage of compressed data, ranging from 32 kbps (for speech) through to 320 kbps. The general default for MP3 encoding is 128 kbps, which is a compromise between file size and quality. If two files are encoded at the same bit rate, one variable and one constant, it will be the recording with the variable bit rate that is generally of higher quality due to the deployment of an increased number of bits for perceptually important information (such as periods of speech) and fewer for less important information (such as periods of low amplitude noise). Variable bit rate encoding also allows a recording to be encoded in multiple passes, such as two-pass encoding, where the first pass is used to analyse the data and determine the areas where the allocation of bits should be highest, but this is generally only used in video.

Interpolation

Interpolation is the creation of new values that lie within a discrete set of data points, based on said points. The term 'inter' indicates the creation of values between two locations within the range of the existing data. The simplest implementation of interpolation is zero-order or previous value interpolation, in which the value of the last sample is held and repeated across the area being created. First-order interpolation or linear order interpolation uses a value derived from the mean value of the previous and subsequent sample values. For interpolation to be possible, some form of *a priori* knowledge is required by the system (Pohlmann, 2005).

Extrapolation

Extrapolation is the creation of values that lie outside of a discrete set of data points, again based on these values. The term 'extra' indicates the creation of values beyond the range of the existing data. For extrapolation to be performed, a model of the process which created the known signal points must first be created before being applied to the region outside the boundary of the discrete data. As with interpolation, knowledge of the discrete dataset is required to allow the expected behaviour of extrapolated samples to be predicted (Fink et al., 2013; Maher, 1994).

Analysis methods

For analysis methods to be understood in their entirety, the most commonly used measurements must first be identified. Methods of analysis can be broken down into three sections based on the domain in which the audio is represented, those of time, frequency, or both.

Time domain

Audio is most commonly represented as a waveform, where time is plotted against amplitude. The x-axis is used to represent the discrete audio samples, although this is often normalised to hours, minutes, or seconds. This allows comparisons to be drawn against recordings of different sample rates and be aligned with a time period with which we are familiar but can result in a reduced resolution in comparison to the raw sample values original captured by the system. The y-axis is used to plot the magnitude of each sample, generally represented as a bipolar value normalised between positive and negative one, but can also be decibels, power spectral density (PSD), or any other magnitude measurement unit.

Transposing the time representation from samples to seconds is simple once the sample rate is known. For example, for a recording with a 44100 Hz sample rate:

Sample No 441 (441/44100) = 1 millisecond

Sample No 44100 (44100/44100) = 1 second

Sample No 2646000 (2646000/44100) = 60 seconds

Another representation of time is that of a 'frames.' This pertains to the splitting of a recording into smaller regions (known as frames), all of an equal number of samples in length. This grouping of samples opens up various types of analysis and potentially faster processing. For example, if a recording consists of 10 samples, when split into frames, each with a length of 2 samples:

1 2 3 4 5 6 7 8 9 10

1 2 / 3 4 / 5 6 / 7 8 / 9 10

In reality, recordings are composed of much larger numbers of samples, and, as such, larger frame lengths are used. Although these lengths are often defined as a number of samples, seconds or milliseconds are sometimes used as it is simple to transpose this to samples as previously shown. One of the drawbacks in using frames is that if a feature (such as a spoken word) falls towards the edge of the frame, and thus half of the said feature is within the current frame and the other half in the next frame, the feature will essentially be lost. This is further compounded when a window function is applied to each frame in the time domain as it emphasises features which fall within the centre of a frame. This is due to the function applying a reduced weighting towards the end of frames, and thus the features in the centre appear more prominently, compared to those at the extremes.

The solution to the potential loss of feature detail is overlapping. This process overlaps consecutive windows based on a user-defined selection, which is commonly a percentage of the frame length. This means that the feature that previously fell across two frames will now fall in the centre of an overlapped frame, and thus is not lost. Percentages are used rather than samples as they provide a comparable point between recordings of different sample rates and durations. For instance, the results of overlapping 100 samples when the frame size is 200 would have drastically different results if the frame sizes were 44100. The most common values are 25 per cent and 50 per cent of the frame length, but any value is technically possible. Although overlapping solves the issues relating to frames, it is not a perfect solution. It too comes with its own problems, as once performed, the data increases in size to the extent that a 50 per cent overlap will result in twice the number of frames needing to be processed (Figure 5.5), and any analysis performed will potentially be performed on the same section of data in different frames (McLoughlin, 2009). Once processing has been applied to the frames, they must be reassembled into a single signal, and concatenation or simply adding the values of overlapped areas is not suitable as it would result in various artefacts, such as clicks at the points at which consecutive frames meet. A final process is therefore required in the form of a window function.

Window functions are used to prevent any edge artefacts and discontinuities by essentially smoothing the data between adjacent frames. Windows are positive, smooth, symmetric functions with a maximum value in the centre that taper to zero at the extremes. They are applied by multiplying the time domain data by the window function. Multiplying by a window with a value of zero at the extremes will smooth the transitions between consecutive windows, subsequently reducing leakage and sidelobes (caused by the restriction of data in the time domain) within the frequency domain (Rust and Donnelly, 2005). Although the Hamming window is recommended as a good starting point, there are several window types available, the most common of which are Rectangular, Hamming, Hann and Blackman. The differences between each relate to their shapes and the degree to which they taper at the extremes.

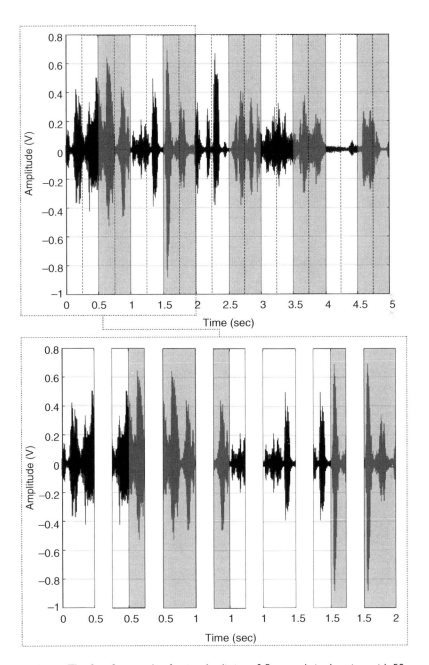

Figure 5.5 Frame theory. The first 2 seconds of a signal split into 0.5 seconds in duration with 50 per cent overlap
Source: Created using Mathworks Matlab.

Power

Power is the total energy emitted within a period of time and is most commonly measured by the SI (International System of Units) derived unit of watts (Thompson and Taylor, 2008). As it applies to audio, it is the total power transmitted from a sound source or the rate at which sound energy leaves the source, and is referred to as 'sound power.' In practice, it is measured in a reverberation chamber or by first measuring sound pressure before calculating the power using these values (Porges, 1977), as sound pressure is directly affected by sound power. A common analogy is to think of sound power as a radiator and sound pressure as the heat within a room.

When plotted, the power of a signal is represented on the y-axis as a function of time or frequency (on the x-axis) and can be represented using a number of different measurements.

Decibels

The decibel is a logarithmic system which was first developed for use in telephony, where the loss of energy from a cable is a logarithmic function of the length. As the human perception of sound is also logarithmic, it was also adapted for use in audio measurements. In order to provide a foundation to this topic, the terms logarithmic and linear should be expanded upon.

Linear scales are those which the general public is most familiar as they form the foundation of the fundamental counting system. The data is simply increased by a constant step size. For example, linear data increasing by a step size of two:

1, 3, 5, 7, 9

Logarithmic scales pertain to those in which there is an increase in step size by a constant proportion.

Logarithmic data increasing by a proportion of two:

1, 2, 4, 8, 16, 32, 64, 128

Although they appear completely different, they are closely related.

The human perception of an increase in amplitude on the linear scale from 1 to 2 would be the same as an increase from 64 to 128 on the logarithmic scale. Both pertain to a single gradation on their respective scales.

The Bel was first proposed by (and named after) Alexander Graham Bell as the log of the ratio between a power to be measured and a known reference power. One Bel represents a change in power by a factor of ten (Hartley, 1928). Thus, three Bels of amplification equates to an increase in power by a factor of 10 x 10 x 10. As the Bel proved to be too large when attempts were made to apply it to smaller measurements (such as sound), it was divided by ten. Hence the 'dec'-ibel (dB) is ten times the log of the power ratio, so three bels of amplification are equal to 30 decibels.

Decibels can be calculated from power using the following equation:

$$dB = 10 \cdot log_{10}\left(\frac{p_1}{p_0}\right)$$

where p_0 represents the power of the reference signal and p_1 represents the power of the signal level to be calculated (Audio Engineering Society, 1955).

Amplitude

Amplitude represents the changes in air pressure relative to the normal atmospheric pressure. When sound travels, the displacement of particles causes regions both higher and lower than this pressure, resulting in a bipolar amplitude (containing both negative and positive values). Voltages are represented in the same manner, evidenced by the peaks and troughs relative to zero in waveform representations of signals. As a microphone uses volts, this is the most common representation of amplitude within digital audio, although it can also be represented in both sound pressure and current. As voltage has a square relationship with electrical power (Ohm's law states $W = V^2/R$), so too does sound pressure to sound power, where, based on the inverse square law, a doubling of distance results in 6 dB of attenuation. Therefore, when calculating decibel ratios, this must be taken into account by squaring the amplitude unit.

$$dB = 10 \cdot log_{10}\left(\frac{(v_1)^2}{(v_0)^2}\right)$$

or more simply:

$$dB = 20 \cdot log_{10}\left(\frac{v_1}{v_0}\right)$$

where 'v_0' represents the reference signal, and 'v_1' represents the signal level to be calculated.

To demonstrate, and determine the dB representation for a doubling in amplitude, imagine a signal of 20 V and a known reference of 10 V. When applied to the formula:

$$dB = 20 \cdot log_{10}\left(\frac{20}{10}\right)$$

$$dB = 6.02 \text{ dB}$$

This shows that a 6.02 dB increase to the level of an audio signal doubles its amplitude. To determine the halving of a signal's amplitude, the positions of the reference and signal to be measured can be swapped. Imagine a signal of 10 V with a reference of 20 V. Applied to the formula:

$$dB = 20 \cdot log_{10} \left(\frac{10}{20} \right)$$

dB = −6.02 dB

As dB is a reference measurement, there are a number of reference standards used within audio, and as the measurement is a ratio, a reference point for comparison is always required. For standard references, this is appended and denoted by a letter appended to the end of 'dB,' the letter being dependent on the reference used (Brixen, 2020). The most commonly found version within audio forensics is that of dBFS (dB Full Scale), and is an option available in the majority of software to represent the amplitude of the signal as a function of time. The reference used for the dBFS measurement is the highest available level within the signal (or Full Scale), which, in an unsigned 16-bit signal is a quantisation level of 65536 (2^{16}). Hence, in order to reduce the signal by 50 per cent, attenuation by 6 dB is required, as demonstrated in the following calculation.

$$20 \cdot log_{10} \left(\frac{32768}{65536} \right) = -6.02 \, dB$$

Other reference values include dBV, in which a reference of 1V RMS is used, and dBm, in which a reference value of 0.78 V RMS is used. These are not generally encountered within audio forensics, but understanding the representations of the power of a signal is of paramount importance for the analysis and enhancement of audio signals.

Normalisation

Normalisation is the process of transposing an entire data set to a specific reference value by dividing the output signal by a given constant (Gonzalez and Reiss, 2008), and can be applied in any domain. For example, during an enhancement, a user might 'normalise' the recording to -1 dBFS. This essentially converts the signal, so -1 dBFS is now the Full-Scale reference value for the signal, and therefore the maximum value for which the signal is represented. This is useful in ensuring the enhancement is at an optimal level without exceeding the maximum quantisation value available to represent amplitude within the system, which would result in clipping.

The most common normalisation is that applied to a waveform representation of a signal normalised to a scale of -1 to 1 (in a signed representation). Here 1 becomes the reference value for positive values and -1 for negative values. The scale is thus converted to exist between the two. This is useful for analysis as it allows a comparison to be made between signals, regardless of their bit depth or magnitude. To normalise the amplitude of a signal, the following formula is applied to every sample within the recording:

sample value x (1/reference value).

The audio can be normalised (or scaled) using the maximum value available within the system to represent amplitude (which will be the maximum quantisation level) as a reference value (known as absolute scaling), or the maximum value of the captured signal (known as relative scaling), depending on the reason for normalisation. In general, absolute scaling is used to ensure the full range of quantisation values available to represent the amplitude of a recording is used, rather than a potentially limited range if the captured signal has a low maximum value. Relative normalisation is used in analysis to compare values within a signal rather than against other recordings.

It should also be noted that normalisation does not improve the SNR in any way, as it is applied to each sample, and thus the entire signal will be transposed, regardless of whether it is desired or not. This means the noise floor will scale in relation to the desired signal, causing it to become more audible in low magnitude signals.

Frequency domain

All processes applied in the frequency domain first require a signal to be converted into its frequency components. Any signal can be represented in either the frequency or time domain and can be converted between the two using a method known as Fourier Transform (and Inverse Fourier Transform respectively), which, due to its applications in everything from music to the study of oceans, it is perhaps the most widely used algorithm in the world today.

Fast Fourier Transform

Fourier Transform is a method of transforming a signal represented in the time domain to a frequency domain representation. The Inverse Discrete Fourier Transform (IDFT) performs the opposite, thus providing a time-domain representation of a signal represented in the frequency domain. The Fourier Series was first proposed and presented in 1807 (Fourier, 1807), although it was not published until some 15 years later due to protests by one of the article's peer reviewers who had a different theory. In his book Fourier (1822) proposed that any signal (x), of any length (t), can be broken down into the individual sine and cosine waves of which it is composed, each having a number of cycles in t. The Fourier Transform can be split into four approaches, based on the properties of a signal, as follows (Smith, 2003):

Fourier Transform: Applied to signals which are *continuous* (extending continuously from negative infinity to positive infinity) and *aperiodic* (without periodic repetition).

Fourier series: Applied to signals which are *continuous* (extending continuously from negative infinity to positive infinity) and *periodic* (repeat in a regular pattern).

Discrete Time Fourier Transform: Applied to signals which are *discrete* (defined at discrete points between negative to positive infinity) and *aperiodic* (without periodic repetition).

Discrete Fourier Transform: Applied to signals which are *discrete* (defined at discrete points between negative to positive infinity) and *periodic* (repeat in a regular pattern).

The first two methods have no application to digital audio as digital data is discrete. Although digital audio data extends between two discrete points, namely, the beginning and the end of the recording, sine and cosine waves are defined from negative to positive infinity. To calculate the Fourier Transform from an aperiodic signal, an infinite number of sinusoids are required, something which is not technically possible. Periodic sinusoids, on the other hand, are by their nature repetitive, and by imagining that the audio signal is just one repetition of a series which extends to both positive and negative and infinity, the criteria are met. The only transform which is therefore applicable to digital audio signal processing is the Discrete Fourier Transform (or DFT), which is both discrete and periodic.

Each approach of the Fourier Transform can also be divided into two versions: real or complex. The complex is, as the name states, complex, and as the technicalities of such are of no use to the average audio forensic examiner, they are outside the scope of this book. Only the real version will, therefore, be considered.

To understand the transform, functions must first be defined. A function is a process that takes an input and produces an output, for example:

A = 2b +5

A transform is a function which allows multiple inputs, so instead of a single input producing a single output, it can take thousands (in the case of digital audio) of samples and produce thousands of outputs.

The number of samples output by the transform does not necessarily have to be equal to the number presented at the input. The Discrete Fourier Transform takes an input signal (the audio samples in the time domain) and produces two frequency domain outputs of $N/2+1$ points, each representing the amplitudes of the sine and cosine waves (where N represents the number of samples at the input). The outputs are called the Real and Imaginary parts, which contain the amplitudes of cosine and sine waves respectively and extend from $0-N/2$. The selection of sine and cosine waves used within DFT each have an amplitude of 1 (unity) and are commonly referred to as the DFT basis functions. The DFT outputs (numbers which represent amplitude) are assigned to the basis functions, resulting in a set of scaled sine and cosine waves of which the original time-domain signal is composed.

Traditionally, the calculation of the Fourier Series was incredibly time-consuming, until a far more efficient algorithm was proposed by Cooley and Tukey (1965) known as Fast Fourier Transform (or FFT). This uses the power of two, where the number of uniformly spaced input samples (N) must be 2^n, as computers use binary coding, making powers of two a natural fit. It, therefore, minimises the number of multiplications, and thus the time to perform the transform. Not all algorithms require N to be a power of 2, but this is the optimal state for the fastest calculations. It also uses complex DFT rather than real DFT, which transforms time-domain signals into two N point frequency-domain signals. The algorithm essentially reduces computations from n^2 to $n(\log)n$, where n is generally an integer.

Although in theory, the calculations result in an infinite number of terms (as the Fourier Transform contains the condition that all sine and cosine waves are considered to be infinite), the only usable frequencies within digital audio are those below the Nyquist rate. As the amplitude values at negative frequencies match those corresponding to positive frequencies, 50 per cent of the frequencies contain all of the information required. This symmetry means that only the output pertaining to $0-N/2$ is required (Donnelly and Rust, 2005).

There are two common ways to display the FFT:

- *amplitude spectrum*, in which the magnitude of the FFT complex values are presented as a function of frequency
- *power spectrum* (also known as a periodogram).

With regards to discrete-time signals, there are two temporal variables to be considered, which affect the resolution of the FFT. The first is the number of input samples, and the second is the number of DFT points (Mathworks, 2018).

Input samples

A good starting point for the number of input samples for an FFT calculation is the number of samples of which the signal itself is composed. The FFT will treat the entire signal as one repetition of an infinite signal of infinite repetitions of the input signal, and calculate the FFT accordingly. This is useful in determining the characteristics of the overall signal, but provides little information in relation to the changing detail within. By splitting the signal into frames, the length of the input signal is limited, resulting in a finer spectral resolution. As the FFT is considered infinite, if the entire signal is treated as s single repetition, the frequencies of a sentence within the middle of an otherwise ambient recording would become lost, and the specific frequencies which make up the sentence would, therefore, be overlooked. As the production of speech is limited by the rate at which our vocal muscles can contract and release, for the purposes of analysis, it can be considered to be stationary for periods of around 20–30 ms. Splitting a signal into frames of 20 ms in length can, therefore, result in the frequencies of which the speech is composed being calculated in isolation, as the transform will now treat the frames as the infinity repeating signal rather than the entire recording.

The maximum frequency that can be represented is, therefore:

$$\Delta f \cdot \frac{N}{2} = \frac{1}{2 \cdot \Delta t}$$

The FFT replicates along the frequency axis with a period of $\frac{1}{\Delta t}$ so, $x(f_{N/2}) = x(f_{-N/2})$. The transform is therefore defined as the interval between $-\frac{1}{2\Delta t}$ to $+\frac{1}{2\Delta t}$, commonly known as the Nyquist band (Rust and Donnelly, 2005).

DFT points/FFT length

The number of DFT points (also known as the FFT length) relates to the number of samples output by the system, which dictates the bandwidth over which the signal is decomposed. This can be calculated using the following methods:

1 Spectral Domain

Spectral Resolution = Fs/n

Where n is the number of DFT points, and Fs is the sampling frequency.
So, for a signal with a sampling frequency of 44100Hz and an FFT length of 1024:

Spectral Resolution = 44100/1024 = 43.07 Hz

The frequency resolution of each FFT component is, therefore, approximately 43 Hz. It, therefore, stands to reason that increasing the FFT length will improve the spectral resolution, which is true, but at the cost of temporal resolution.

2 Temporal Domain

Temporal Resolution = n/Fs

So, for the above signal specifications:

Temporal Resolution = 1024/44100

Temporal Resolution = 0.0232 s, or 23.2 ms

This means the FFT will be calculated every 23.2 milliseconds, so signals shorter than this will not be represented once they have been transformed to the frequency domain, just as a sentence within the middle of an ambient recording would not be considered represented if the FFT input length was equal to the total number of samples within a recording.

As the resolution of the frequency domain is the inverse of the time domain, both cannot be improved simultaneously. Thus, a balance must be sought between the spectral resolution and the temporal resolution, and considerations made as to the type of signal being analysed and the reason for the analysis through the careful selection of the length of the input and the length of the FFT output. With that being said, there is a method to improve the representation of the frequency domain by adding zero value samples to the time domain, thus, increasing the number of data points for the same frequency range (through interpolation), and performing the same in the frequency domain to increase the time resolution of the data.

Zero padding

Zero padding is the appending of a string of zeros to the beginning, end, or the middle of the data (the selection of their placement does not affect the outcome). Zero padding is applied to FFT for two reasons:

1   to ensure the data in the time domain is a power of two, and thus the calculation is more efficient;

2   to improve the interpolation in the transformed domain.

Although it may appear that dividing the input into smaller frames and applying an FFT which matches the length of each frame will ensure the maximum resolution is obtained, this is not the case. While this sounds good in theory, there is another issue relating to the use of windows. When splitting the signal into frames, the result is a rectangular window function, that is, there is no consideration for the frame edges and the effect these have on the FFT. If you imagine using the entire length of the signal as an input, it would be considered a rectangular window as it is treated by the FFT as a single repetition of an infinitely repeating signal. The result of this rectangular window function is the presence of artefacts in the form of side lobes at either side of the main lobe. When a window function is plotted, the main lobe is centred at each frequency component of the time-domain signal showing the frequency characteristics of said window. Lower sidelobes reduce leakage into the frequency being measured but increase the bandwidth of the main lobe (Mathworks, 2018).

Power spectral density (PSD)

Audio signals do not have finite energy from sample to sample, and as such, the associated variations make it impossible to determine their future. They do, however, have finite average power, which can be characterised by the average power spectral density (Stoica and Moses, 2005).

Power spectral density estimates a signal's energy (or strength of variations) as a function of frequency, and in doing so shows the frequencies for which the variations are strongest and those which are weakest across the spectrum. Only an estimation is possible as the spectra of finite data suffer from poor resolution and leakage caused by non-zero spectral components at frequencies other than harmonics due to sampling over non-integer multiples of the signal period (Mathworks, 2018).

It is best understood by breaking the term down into the elements of which it is composed. The first, 'power,' defines the power, not in the traditional sense, but calculated from the mean square value of the signal at a given point in time. The mean square value is used, as calculating the mean of a signal at a specified point in time will take both positive and negative frequencies into account, resulting in a value close to zero (otherwise known as the DC Offset). By squaring values, negative numbers are transposed to positive ones. Once this has been performed, the power can then be calculated. 'Spectral' refers to the spectrum, thus indicating it is a function of frequency rather than time. 'Density' refers to the fact that it is calculated by splitting the signal into smaller bandwidths and measuring the normalised value of such (thus allowing comparisons to be drawn between bandwidths), and thus the density across the spectrum.

Power spectral density can, therefore, be summarised as the distribution of power across the frequency spectrum (Figure 5.6).

There are various methods to estimate the PSD, but those used in audio forensics are classified as non-parametric as they relate to the signal itself, with no respect for the capture. To calculate the PSD, the power is divided by the bandwidth from which it relates. If normalisation is not

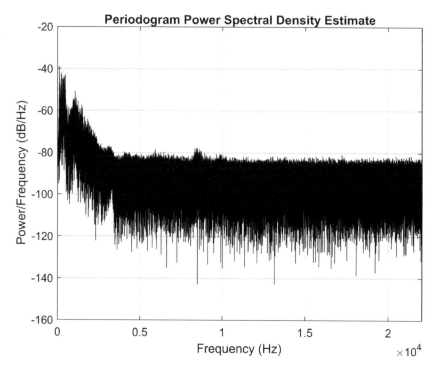

Figure 5.6 Power spectral density of an uncompressed audio signal
Source: Created using Mathworks Matlab.

performed, it would not be possible to compare the measurement of each bandwidth as the distribution is highly dependent on the number of bands over which the audio is split (Siemens, 2019).

The measurement unit is not standardised as the method can be applied to a number of areas within digital signal processing, but for forensic audio purposes, it is generally presented in dB/Hz. The power must, therefore, be converted to decibels using the method covered earlier in this chapter.

It can be presented to provide the energy per frequency band across the entire signal, or the signal can be framed to show the energy in the time domain (Figure 5.7). It is also possible to determine the relationship between two signals through the use of the cross spectral density (CSD) function, which computes the cross-correlation of such.

LTAS

LTAS stands for long-term average spectrum. This is a representation of the mean frequency distribution of the signal's energy over a long-term period, so as not to be peaked by individual events and transients which would appear within a shorter FFT sample size. It provides an overview of the frequency content of the entire audio signal by taking the average of large frame sizes

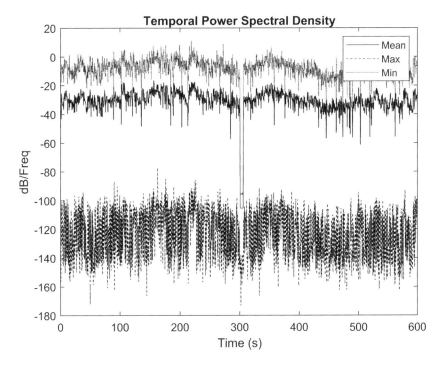

Figure 5.7 Windowed power spectral density containing a signal loss event at approximately 300 seconds
Source: Created using Mathworks Matlab.

and is generally measured in dBFS (Leino, 2009; Sundberg et al., 1987). The value used for the framing of the signal is for the user to decide, but 1 second is a good starting point.

Ordered

Both the PSD and LTAS can be represented in an 'ordered' format, where they are called the ordered PSD and LTASS (long-term average sorted spectrum) respectively. This process orders (or sorts) the data in descending energy values to provide a tempered representation of the energy of the signal (Grigoras et al., 2012).

Summary

Concepts pertinent to audio forensics encompass aspects of acoustics, electronics, and digital signal processing. Although these concepts are relevant to both recording studios and forensic examiners alike, the differences in the tasks performed by each dictate the level of understanding required. While recording studio engineers may focus on microphone placement concepts and theory, this is generally outside of the forensic examiner's control so it may be an area for which they have less understanding. Only an overview of topics relating to tasks which forensic examiners would be expected to encounter have been presented, and the reader is encouraged to review further topics and details to broaden their understanding of the audio world.

References

Audio Engineering Society, 1955. Decibel tables. The Journal of the Audio Engineering Society 3.

Audio Engineering Society, 2018. Preferred sampling frequencies for applications employing pulse-code modulation (No. AES5-2018). Available at: www.aes.org/publications/standards/preview.cfm?ID=14

Barrow, B., 1997. A lesson in megabytes. IEEE Standards Bearer January.

BBC Radio 4, 2008. The man who invented stereo. Archive Hour.

Brixen, E.B., 2020. Audio metering: Measurements, standards and practice, 3rd edn. Routledge, Abingdon.

Cooley, J.W. and Tukey, J.W., 1965. An algorithm for the machine calculation of complex Fourier series. Mathematics of Computation 19, 297–301.

Donnelly, D. and Rust, B., 2005. The Fast Fourier Transform for experimentalists, part I: Concepts. Computing in Science and Engineering 7, 80–88.

Fink, M., Holters, M., Zölzer, U., 2013. Comparison of various predictors for audio extrapolation. Paper presented at the 16th International Conference on Digital Audio Effects, Maynooth, Ireland.

Fourier, J., 1807. Mémoire sur la propagation de la chaleur dans les corps solides. L'Institut national. Nouveau Bulletin des sciences par la Société philomatique de Paris.

Fourier, J.B.J., 1822. Théorie analytique de la chaleur. Chez Firmin Didot.

Gonzalez, E.P. and Reiss, J., 2008. An automatic maximum gain normalization technique with applications to audio mixing. Paper presented at the 124th Convention, Audio Engineering Society, Amsterdam, The Netherlands.

Grigoras, C., Rappaport, D., Smith, J.M., 2012. Analytical framework for digital audio authentication. Paper presented at 46th AES International Conference: Audio Forensics, Audio Engineering Society, Denver, CO, USA.

Hartley, R.V.L., 1928. 'TU' becomes 'Decibel' (No. 4), Bell Laboratories Record.

Huber, D.M., Runstein, R.E., 2005. Modern recording techniques, 6th edn. Elsevier, Oxford.

Jayant, N.S. and Rabiner, L.R., 1972. The application of dither to the quantization of speech signals. The Bell System Technical Journal 51, 1293–1304.

Katz, B., 2002. Mastering audio: The art and the science. Focal Press, New York.

Koenig, B.E., Lacey, D.S., and Killion, S.A., 2007. Forensic enhancement of digital audio recordings. Journal of the Audio Engineering Society 55, 352–371.

Leino, T., 2009. Long-term average spectrum in screening of voice quality in speech: Untrained male university students. Journal of Voice 23, 671–676.

Letowski, T.R., 1992. Anchor effect in an optimum timbre adjustment. Journal of the Audio Engineering Society 40, 706–710.

Maher, R.C., 1992. On the nature of granulation noise in uniform quantization systems. Journal of the Audio Engineering Society 40, 12–20.

Maher, R.C., 1994. A method for extrapolation of missing digital audio data. The Journal of the Audio Engineering Society 42, 350–357.

Mathworks, 2018. Signal processing in Matlab. Available at: www.mathworks.com/products/signal.html

McLoughlin, I. 2009. Matlab for speech and audio processing. Cambridge University Press, Cambridge.

Nyquist, H., 1928. Certain topics in telegraph transmission theory. Transactions of the American Institute of Electrical Engineering 47, 617–644.

Painter, T. and Spanias, A., 2000. Perceptual coding of digital audio. Proceedings of the IEEE 88, 451–515.

Pohlmann, K.C., 2005. Principles of digital audio, 5th edn. McGraw-Hill, New York.

Porges, G., 1977. Applied acoustics. Wiley, London.

Rossing, T.D., Moore, R.F., and Wheeler, P.A., 2013. The science of sound, 3rd edn. Pearson, Harlow.

Rust, B., Donnelly, D., 2005. The Fast Fourier Transform for experimentalists part III: Classical spectral analysis. Computing in Science and Engineering 7, 74–78.

Siemens, 2019. What is a power spectral density? Available at: community.sw.siemens.com/s/article/what-is-a-power-spectral-density-psd

Smith, S.W., 2003. Digital signal processing. Elsevier, Oxford.

Sreenivas, T.V. and Dietz, M., 1998. Vector quantization of scale factors in advanced audio coder (AAC). Paper presented at the 1998 IEEE International Conference on Acoustics, Speech, and Signal Processing, IEEE, Seattle, WA, USA.

Stoica, P. and Moses, R.L., 2005. Spectral analysis of signals. Pearson/Prentice Hall, Upper Saddle River, NJ.

Sundberg, J., Ternstrom, S., Perkins, W.H., and Gramming, P., 1987. Long-term-average spectrum analysis of phonatory effects of noise and filtered auditory feedback. STL-QPSR 28, 57–80.

Thompson, A. and Taylor, B.N., 2008. Guide for the use of the international system of units (SI). NIST, Gathersburg, MD.

Zjalic, J., 2018. Why forensic audio isn't audio engineering. Sound on Sound. Available at: www.soundonsound.com/techniques/why-forensic-audio-isnt-audio-engineering

Chapter 6

Audio encoding

Introduction

Once sound has been sampled, it must then be encoded in the form of a digital file for storage and subsequent playback. Although there are many audio formats, with each differing from the next, the general components and principles are the same for all digital audio files. These are most easily understood when deconstructed into several distinct areas. Understanding how audio is encoded is essential to a number of forensic processes, from conversion through to authentication.

There are several elements which make up a digital audio file, all of which pertain to either the file container or the audio data. The following paragraphs provide a brief overview of these elements and how they all relate to one another.

A *codec* (taken from a portmanteau of the words *code* and *decode*) is a software algorithm installed on a computer operating system, capable of encoding and decoding multimedia data, of which a number of varieties exist. During the encoding stage, the raw audio data is processed by an algorithm to reduce the size of the data and thus aid in smaller file sizes. This process is essential to increase the number of recordings that can be stored on a digital medium and for streaming at lower bandwidths. When decoded, this data is processed to represent the original audio data. The extent of the similarity between the original and decoded version is dependent on both the type and degree of compression applied by the algorithm.

When an audio file is saved within a file system, the encoded audio data (the stream/payload) is stored within a container or wrapper, with a filename extension associated with the container. The filename extension is used by the operating system to determine the software required to read the structure of the file and play the recording. The container holds information in relation to the codec used to encode the data (and thus the codec required to decode the data), information required by the decoder to decode the data and all metadata associated with the file (Figure 6.1). The file itself does not contain any form of codec. MP3 encoded files do not require a file container.

Audio and video playback software contain various decoders to ensure they can read audio files encoded with a wide variety of codecs. In order to do so, they often contain codec libraries to store the most common codecs available. These libraries are often created by a third party company and then used by software developers, and are then installed in addition to the pre-existing codecs which are shipped with the software. For example, FFmpeg (Fast Forward MPEG) (SWGDE, 2017) is the developer of 'libavcodec,' an extensive open-source library of

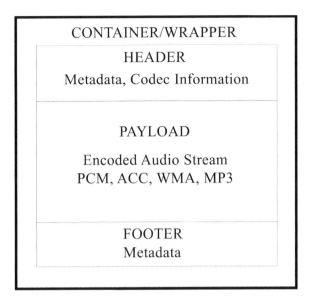

Figure 6.1 Structure of an audio file

over one hundred audio and video codecs used by both commercial software and freeware (FFmpeg, n.d.).

Confusingly, the term 'format' can be used to refer to either the container or the codec used but is generally used to describe the container format.

Filename

The filename of the audio is self-explanatory, and can be defined by a user, the capture device, or by the capture device and then changed by a user. It is clearly visible within the file structure of an operating system with no requirement for specific software, and its default format, when saved by the capture device, will be dependent on said device, as there is no standard. Some people use the start or end time of the capture, whereas others use sequential numbering such as recording 1, recording 2, etc. Some of the systems which are GPS-enabled also include the location within the filename.

The filename itself is of little importance as it has no influence on the contents of an audio file, and is extremely straightforward to manipulate. It would not, therefore, be relied upon in any way during an authentication examination, although it may provide information in support of other findings.

Filename extension

The filename extension consists of the three or four characters of text which appear after the filename to represent the file container (which may require the selection of a specific option with

Table 6.1 Common audio filename extensions and their formats

Extension	Standard	Format	Metadata tag
.aac	International standard	MPEG-2 Part 7 AAC format.	No, but can use ID3
.aiff	Apple standard	Apple Audio Interchange File Format.	The basic form has no metadata tag
.flac	Xiph.org open standard	Free lossless Audio Codec file in OGG container.	Yes
.mp3	International standard	No container, uses MPEG 1 Layer 3 format	No, but ID3 can be added
.mp4	International standard	Can contain a variety of audio and video coded data, most commonly AAC.	Yes
.m4a	International standard	MP4 container but with audio content only.	Yes
.ogg	Xiph.org open standard	Vorbis file in an OGG container.	Yes
.opus	Xiph.org open standard	Opus audio file in an OGG container.	Yes
.wav	IBM/Microsoft standard	Uses RIFF container. Can contain a variety of formats but usually uncompressed audio data.	INFO chunk, but ID3 and XMP can be added.

some operating systems for it to be visible). It is used by the operating system to determine the software needed to open the file, but if the software is able to open a file container (based on the extension), that does necessarily mean it has the decoder installed to play the audio data. In this case, it will likely not play or may cause the software to crash. Table 6.1 presents some of the most common audio filename extensions and their associated formats.

File structure

To view the data in a format that allows the structure to be both visible and understood, a hex viewer/reader software is required. There are numerous options on the market, both freeware and commercial, all of which allow the structure of the file to be visible. However, some do have other features such as tools to search through the structure for keywords a user has entered and provide colour-coded indicators to show the location of various important elements. The choice of hex viewer software used is down to personal preference.

The term 'hex' comes from the word *hex*adecimal, which is a numeric system that presents binary computer data in a format that can be comprehended by humans. For it to be in a language we understand, the hexadecimal data must then be converted to a human-readable format such as ASCII (American Standard Code for Information Interchange). Just as binary can be converted to ASCII, it can also be converted in the opposite direction from ASCII to binary, thus allowing humans (and more specifically software developers) to include data such as the make and model of the device, or the date and time at which the audio was encoded (Figure 6.2).

```
Offset(h)  00 01 02 03 04 05 06 07 08 09 0A 0B 0C 0D 0E 0F   Decoded text

00000000   00 00 00 1C 66 74 79 70 4D 34 41 20 00 00 00 00   ....ftypM4A ....
00000010   4D 34 41 20 6D 70 34 32 69 73 6F 6D 00 00 00 01   M4A mp42isom....
00000020   6D 64 61 74 00 00 00 00 00 0E 7C 03 00 D0 40 07   mdat......|..Ð@.
00000030   00 F8 99 CC 3D BC D0 54 94 CC D0 81 69 8B 52 BB   .øᵐÌ=¼ÐT"ÌD.i‹R»
00000040   F7 F9 FE 59 23 CB 44 3A F0 2F 0F 80 81 0D 9D 72   ÷ùþY#ËD:ð/.€...r
00000050   6C 59 05 30 98 03 8F 03 8F C1 EB 1A 4C 16 E0 91   lY.0˜....Áë.L.à'
00000060   34 41 FF 1E C8 CC B3 B8 E3 B2 59 61 61 41 63 38   4Aÿ.ÈÌ³¸ã²YaaAc8
00000070   1E F8 FE 83 FC E8 04 E1 BF 0A 4E 9E BD 42 16 F3   .øþ.üè.á¿.Nž½B.ó
00000080   0D 14 1D 6B D1 E2 C1 B4 AA 64 16 DE 69 82 24 CE   ...kÑâÁ´ªd.Þi.$Î
00000090   D4 D8 67 7B A2 3C 30 1D 44 ED 62 C0 21 49 D6 2A   ÔØg{¢<0.Díbà!IÖ*
000000A0   CE E2 53 C0 DC 80 F5 44 E8 15 BD A8 5C 7C CA CE   ÎâSÀÜ€õDè.½¨\|ÊÎ
000000B0   01 2A D4 AC 8C F4 49 1C 46 03 1E 74 B5 AE CD A1   .*Ô¬Œôî.F..tµ®Í¡
000000C0   CD 80 2D 0C B0 B3 91 52 9E 39 8A AA A5 59 3B A8   Í€-.°³'Rž9Š²¥Y;¨
000000D0   E9 9F 8D 2E D7 55 76 E9 2F 11 EB F1 6A 1B E9 D9   éŸ..×Uvé/.ëñj.éÙ
000000E0   A1 47 0D A3 21 95 56 02 4E DD B3 39 62 2C B2 DF   ¡G.£!•V.NÝ³9b,²ß
000000F0   5B 2B 66 0D 38 6F 5D 42 6D 60 14 59 BE EC D5 3C   [+f.8o]Bm`.Y¾ìÕ<
00000100   38 79 F0 B4 92 04 28 9A A2 B9 E7 ED EA E3 0B B8   8yð´'.(š¢¹çíêã.¸
```

Figure 6.2 Hexadecimal/ACSII representation of a section of an MPEG-4 audio file
Source: Captured from HxD.

Container

Although, in its rawest form, audio data can be stored in a manner completely representative of the stream sampled and quantised (such as PCM), it is usually stored within a container.

A file container or wrapper (as it wraps around the actual audio data or 'payload') is used to combine different data types of data in a single file. There are various types of containers available, and some can contain just audio, whereas others can contain both video and audio streams, as well as other information such as subtitles and metadata.

Essentially, everything which is not payload data representing the quantised audio, including format specifications, encoding-related information, and any information the software manufacturer decides to include, is stored within the container.

Upon playback, the software then uses the information stored within the container to determine factors, such as the decoder required for reproduction and the location within the file where the audio data actually begins and ends. A hex editor/viewer can be used to view the content of a container, which exists as a header (at the start of the file, preceding the payload) and a footer (at the end of the file, trailing the payload). The size of the header and footer is not standardised, so can vary from a few to thousands of bytes.

The elements which define the structure of the data have different names, dependent on the container type used. For example, the Quicktime/MP4 format elements are called atoms (Apple Computer, Inc, 2016), while the RIFF format elements are referred to as chunks (Gibbs, 1999). Most consist of these elements in a sequence, each with an identification header. This prevents the corruption of the entire audio file if a single chunk is damaged, and allows decoders to be standardised to ensure playback on all systems. There are a few containers specific to only audio, such as AIFF (audio interchange file format), a specification based on the IFF (interchange file format), and WAVE (Windows audio file format), based on the RIFF (resource interchange file format) specification. More sophisticated containers allow streams of a range of content to be stored, such as 3GP (commonly used by mobile phones and based on an ISO base media file format), AVI (the standard Windows container, also based on the RIFF format), MPEG (used to store MPEG-1 and MPEG-2 format files) and MP4 (the standard for MPEG-4 multimedia, based on ISO media file format).

Audio data

Between the header and the footer exists the audio data, also known as the 'payload' (Kaur and Dutta, 2018), which, again, can be viewed using a hex editor/viewer. The size of this section will be dependent on the encoding method used and the duration of the audio capture. The storage of the data is also dependent on the encoding method. For example, WAV PCM files store the samples linearly, whereas MP3 files store the encoded data (which is not representative of samples due to the lossy compression process) in frames, each of which can be located by an identifier at the beginning of each frame. Although there may be some readable data in the form of ASCII converted text with regards to frame headers, the vast majority of the time the payload data will not be legible.

Uncompressed audio

PCM

PCM (pulse code modulation) refers to a method in which a digital representation of an analogue signal is reconstructed through discrete amplitude and time domain sampling. First conceived by Alec H. Reeves in 1937, it was initially developed to transmit digital signals over analogue communication channels (Reeves, 1965) and exists in various forms, including LPCM (linear PCM), A-law, u-law, DPCM (differential PCM) and ADPCM (adaptive differential PCM). Although there are various encoding methods, the PCM term has come to refer to the linear format unless otherwise indicated, in which quantisation levels are uniformly distributed across the range of values used to represent amplitude data. LPCM is used in a number of standard digital audio formats, including WAVE, AIFF, DVD audio, the Redbook Standard (used by the audio CD) and transmission through HDMI. The name pulse code modulation stems from the representation of sampled values as digital pulses for transmission and storage (Jayant, 1974).

The steps for the encoding of LPCM data are as follows:

1   A band-limited signal is sampled at the Nyquist rate to prevent aliasing.
2   The amplitude of the signal is quantised to 2^B levels, where B represents the number of bits per sample (also known as the bit depth).
3   The now discrete amplitude levels are represented by binary words of length B. For example, if B = 2, there would be four levels available: 00, 01, 10, 11.

4   The binary words are represented as a series of modulated pulses signifying the amplitude at sample times.

Upon decoding, the process is reversed. The pulses are demodulated and error corrected, and as the pulse codes represent binary words, they can be simply mapped back to amplitude levels before the sequence is low pass filtered to avoid aliasing. As only the presence or absence of a pulse is required to read the signal, this method is extremely robust and allows a signal to be reconstructed with no loss (Pohlmann, 2005).

Delta modulation

Delta modulation is a simplified form of pulse code modulation. It employs an extremely high sample rate, using a 1-bit quantiser to represent an amplitude increase or decrease relative to the previous sample. If the amplitude increases, a value of one is stored. If it drops, a value of zero is stored. Adaptive delta modulation is similar, but with a quantisation step size that adapts, based on the direction the amplitude is moving. The step size increases when quantiser overload is detected (when the number of quantisation levels available is too coarse to represent the input) and decreases during underload. In simpler terms, the step sizes become gradually smaller until consecutive bits are moving in the same direction (increasing or decreasing in amplitude) at which point the step size increases (Aziz et al., 1996).

ADPCM

ADPCM (adaptive differential pulse code modulation) combines the adaptive delta modulation technique with pulse code modulation by making instantaneous changes to step size. It was developed by Bell Labs in the early 1970s and was designed for the coding of speech, or more specifically for digital telephone applications and the storage of digital speech (Cummiskey et al., 1973). It uses a bit rate of just 24–32 kb/s, so occupies half the bandwidth of PCM by removing redundant data from the signal. This results in a greater dynamic range, evidenced by the reduced level of noise during quiet sections. It also uses the same adaptive quantiser step size technique as adaptive delta modulation but produces speech of a perceptually higher quality. Rather than encoding the signal itself, ADPCM represents the prediction error, calculated by subtracting the difference between the original signal and a prediction of the signal based on the previous behaviour of the signal. It must, therefore, be decoded through a reverse transform in order to reconstruct the signal.

Although ADPCM requires a linear PCM input signal, a logarithmic quantiser is used by telephony to achieve reductions in bandwidth, due to the non-linear nature of speech. U-law (pronounced mu-law) was adopted by the US, Canada, and Japan and takes a 14-bit linear input, while a 13 bit-linear input system of a-law was implemented by the rest of the world (which is scaled to 14 bits from 13). The a-law and u-law encoding methods are part of the international standard G.711. Upon transmission, these signals are mapped to 8-bits (Dimino and Parladori, 1995; Texas Instruments, 2002).

This coding method is most commonly found within VOIP (Voice Over Internet Protocol) applications that focus on speech and the digital storage of such, for example, call centres, including both the emergency services and commercial organisations.

WAVE

The WAVE (waveform audio file format) format is based on the RIFF (resource interchange file format), a file structure for the storage of multimedia data, and one which encompasses more specific file formats including AVI. It was developed by Microsoft and IBM, and first released in 1991 (Microsoft Corporation and IBM Corporation, 1991). To specify the format, a 'WAVE' offset follows directly after the RIFF chunk ID within the 12-byte header structure of such files. The term 'chunk' is used by the RIFF specification to define elements of the file structure. Other formats use different terminology, such as 'atoms' in the Apple Quicktime format. There are only two mandatory chunks within a WAVE file structure, 'fmt' and 'data,' and the 'fmt' chunk must always precede the 'data' chunk. The format chunk is positioned after the file header and contains the parameters of the audio, such as the sample rate, channel structure, and bit depth. The 'data' chunk contains all of the actual audio data (Koenig and Lacey, 2014). There are several WAV formats, for example, Microsoft WAVE PCM, IBM mu-law, and IBM a-law (used as a European telephony standard) (Pope and van Rossum, 1995).

Uncompressed WAV PCM is one of the highest quality specifications available, as the audio data stored represents the signal directly after quantisation, without any form of compression. In these recordings, the quantised sample values are, therefore, stored as integers within the payload section of a file. In a monaural recording, they are stored consecutively, such as:

M, M, M, M, M, M

In a stereo recording, the samples are interleaved, such as:

L, R, L, R, L, R

Each sample represents a 'sample point.' Within stereo recordings, the samples pertaining to the left and right channels (which are replayed simultaneously on playback) are classed as 'sample frames.' The sample frame for mono recordings is a single sample point.

Regardless of the bit-depth, the samples are stored linearly, but there are differences between their representation when quantised to 8 bits as opposed to 9 bits or above. 8-bit quantisation values are stored as unsigned integers (so no polarity sign such as plus or minus), with a minimum value of 0 and a maximum value of 255.

For audio stored using quantisation values of 9 bits or above, the samples are stored as signed integers, so extend from negative to positive values around a central point of zero. The LSB (least significant bit) is stored first, followed by the MSB (most significant bit), and the remaining bits set to zero. The LSB is the bit in a binary sequence which carries the lowest numerical value and therefore has the least effect on the value of a binary number. The MSB is the highest numerical value and has the most significant impact on the value of the binary number.

If compressed, each sample can be stored using a differing number of bytes (Microsoft Corporation and IBM Corporation, 1991).

AIFF

AIFF (audio interchange file format) is the Apple Corporation's equivalent of WAVE, released some three years earlier as the result of ten months of meetings between Apple and music developers during 1987 and 1988. It is not as popular as WAVE, possibly due to the commonality between the WAVE format and the audio CD, both of which use the same LPCM method to store data, but have different standards (CDs use the Red book IEC 60908 standard). AIFF conforms to the EA IFF 85 Standard for Interchange Format Files developed by Electronic Arts and also stores audio data in a PCM format (Apple Computer, Inc, 1989).

Compressed audio

Although the previously described formats encode audio in a manner which results in an accurate representation of the analogue signal at capture, the amount of data required to store such information can be substantial. In order to reduce the file size, and thus increase the capacity for storage and transmission, a technique known as compression is used. The theory behind this method was first proposed by C.E. Shannon in 1948, in which the limitations of data compression without losing any quality (otherwise known as lossless compression) were demonstrated. It was also proposed that codes could be used to compress information, in what is commonly referred to as Shannon Entropy. Lossless compression reduces the file size for storage during the encoding stage, but, upon decoding, the file is reconstructed without any data loss.

Lossy compression reduces the file size as above, but upon decoding, any data removed during encoding is not recovered. This results in a degree of quality degradation, the tolerance of which is based on the codec used and the settings of such. It is performed through various processes which take advantage of areas of deficiency in the human auditory system, such as critical masking and bandwidth reduction. Lossy compression produces the smallest audio file sizes available, thus explaining its popularity with streaming services and applications where storage is at a premium.

Rather than reinventing the wheel, the vast majority of audio formats are based on published standards, and codecs are created by various companies and organisations to comply with these standards. This ensures global compatibility, regardless of the software used for playback.

Perceptual encoding

To reduce the data size of an audio file by a significant extent, some data must be removed. 'Perceptual encoding' is an umbrella term used to classify a variety of techniques, all of which take advantage of the theory that humans cannot perceive all of the data being produced on playback of PCM recordings. The entire design of perceptual encoders is focused on economy, and one of the most straightforward techniques is to remove the frequencies which we cannot perceive, or those for which removal would not have a noticeable impact. As discussed in Chapter 4, the broadest frequency range over which humans can hear is limited to between approximately

0–20 kHz. As we age, this is reduced, and even people who can perceive audio up to 20 kHz have a poor perception in the uppermost frequencies, due to decreased sensitivity in these regions.

The other techniques take advantage of the effects of masking, which is defined as either simultaneous or non-simultaneous. Simultaneous (or critical) masking refers to two or more sounds composed of similar frequencies which occur at the same time, but due to the frequency resolution of our hearing, we cannot differentiate between these frequencies due to their proximity to one another. Research by Zwicker (1961) proposed that human perception within the frequency domain can be separated into 24 critical bands. Further research found that sounds composed of frequencies which occupy separate bandwidths can be distinguished as two distinct sounds, but if they occur within a single band, they cannot (Fletcher, 1940).

Non-simultaneous masking relates to audio which does not occur at the same time but does so within an extremely close temporal proximity (Deatherage and Evans, 1969). Pre-masking relates to audio which appears within 50 ms of the start of the 'masked component,' and post-masking is caused by audio which occurs within 50–300 ms of the signal ending. During this type of masking, the stronger component will mask the weaker one.

A key component in the success of perceptual encoding is, therefore, based on masking, and more precisely critical bands, which can be modelled as a series of overlapping, nonlinear filters known as a filter bank. This non-uniform band splitting is composed of constant bandwidths filters up to approximately 500 Hz, and filters with a constant quality factor (or 'Q' factor) above this. As higher frequencies are perceived less by the human auditory system, and also generally require more space for encoding, it makes sense that these areas are coded with the lowest number of bits.

The level at which masking occurs is called the masking threshold, and it varies over time, dependent on the amplitude, frequency, and character of the 'masker' (the sound causing the masking) and the 'maskee' (the sound being masked). Higher frequencies are more susceptible to masking, again due to our reduced sensitivity in that area. The masking threshold is also smaller for noise, indicating noise is a better masker. A signal-to-mask ratio is therefore calculated from the masking threshold to determine the amount of noise that can be masked by an input signal. This signal-to-mask ratio is then applied during encoding to ensure the quantisation noise remains below the masking threshold.

Although the techniques used are all based on a single model (Figure 6.3), a variety of formats are available, each with their own application of the general process.

MP3

MP3 is probably the best-known and ubiquitous audio file format in the world and was born out of necessity for smaller digital audio file sizes. In the late 1980s, the International Organization for Standardization (ISO) declared that there was a need for a digital format of reduced file size without a significant compromise on quality in comparison to the uncompressed formats available at the time. ISO is an international organisation which aims to facilitate the collaboration of countries through published standards for all to follow. Interestingly, the abbreviated term is Latin for 'equal,' hence 'ISO' rather than what should be 'IOS' if abbreviated in accordance with the acronym. The task of developing standards for the compression of both video and audio data was assigned to a working group within ISO, known as the Motion Picture Experts Group

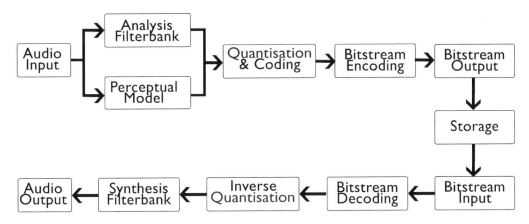

Figure 6.3 Overview of the perceptual encoding and decoding process

(MPEG). An essential component of the standardisation is that any codec which complied with the standard can decode data encoded by another compliant codec.

The development began in 1988, and in 1992, the MPEG-1 standard was announced. As it relates to audio, the group (chaired by H.G. Musmann of the University of Hannover) proposed three methods which increased in complexity, entitled layer I, layer II, and layer III respectively (Musmann, 2006). It is the third, or more officially, MPEG Version 1, Layer 3 (more commonly shortened to MP3) which has become the most popular, due to its efficiency, significant compression capabilities (12:1 ratio), and a tolerable degree of quality degradation for the applications which it is used. MP3 files do not use a standardised container but exist as a sequence of independent, self-contained frames to which tags are appended to store information (Roussev and Quates, 2013).

It was actually the Fraunhofer Society (a German research organisation) in collaboration with the University of Erlangen (and financed by the EU) that first started working on a method of low bit rate audio encoding. It was patented in Germany in 1989 before being submitted to ISO, who then published the specification as part of the MPEG-1 standard in 1992. It was not until 1996 that Fraunhofer received a patent for the MP3 specification in the US (Hacker, 2000).

Although the first MP3 release was effective at higher bit-rates, its capability for encoding high-frequency content was poor, leading to the development of 'spectral band replication' technology, or SBR. This technique was crucial for generating the high-frequency components and was released in a later version called MP3Pro (Ziegler et al., 2002).

Although defined by ISO, a large amount of freedom is given as to the implementation of an MP3 codec and the associated specifications. Those which can be defined by the user will be discussed first.

Bit rate

The bit-rate can be between 8–320 kb/s, although the default is 128 kb/s. This rate is not per channel (if the recording contains more than a single channel) but per recording. So for stereo recordings, the number of bits will be distributed based on the complexity (and thus the data requirements) of each channel at each moment in time within the recording.

There are two types of bit rate available: constant bit-rate (CBR) and variable bit-rate (VBR). Constant is precisely as the name implies, as every section of the recording is encoded with the same number of bits, regardless of the complexity of the audio at that period. Variable distributes the bit rate based on the dynamics at a set period within the recording. The quality of VBR recordings is definable during encoding by the user as the 'maximum bit rate,' but it is usually the average that is provided upon viewing the specifications once decoded (Raissi, 2002).

Sample rate

MP3 supports sample rates of 32 kHz, 44.1 kHz, and 48 kHz (Gadegast, 1996).

Channel modes

There are four different channel modes available for MP3 recordings to be stored:

- single channel (monaural);
- dual channel (two independently encoded channels or double mono);
- true stereo;
- joint-stereo (M/S and intensity encoding).

As the first three have been covered in Chapter 5, only the joint-stereo techniques will be discussed further.

- *M/S (mid-side)*: This is a lossless perceptual encoding method for the storage of stereo channel information to reduce the bandwidth or storage required by a digital audio file. Upon encoding, the left and right channels are summed to create an 'M' or mid channel, while the difference between the left and right channels is calculated to create an 'S' or side channel. Upon decoding, the previously summed 'M' channel is added to the 'S' channel to reconstruct the left channel, while the 'S' channel is subtracted from the 'M' channel to reconstruct the right. This method is highly economical as the sum of two channels require the same amount of data as a single channel, and the difference channel ('S') will require fewer storage bits than a single channel, due to the reduced amount of data it needs to store. This results in one entire channel ('M') and a smaller channel ('S') being stored rather than two entire channels. This method is most efficient when two channels are highly correlated as the difference signal will not require much data, and, as such, it allows a more substantial degree of compression than transmitting the channels independently (Doliwa, n.d.; Herre and Eberlein, 1992).
- *Intensity*: The spatial perception of sound by the human auditory system is dependent on frequencies, and whereas the spatial perception of low-frequency sounds requires the magnitude and phase of both stereo signals to be evaluated, the perception of higher frequencies relies on the energy-time envelopes. Intensity coding takes advantage of this by only transmitting a

single channel of spectral values in relation to the upper-frequency range, which can be shared by the two channels upon decoding. The original energy-time envelopes are preserved through the use of a scaling operation, so each channel's signal is then reconstructed to its original level when decoded. The most straightforward implementation of this is through the use of the two-channel concept of MS coding, in which one contains the spectral values and the other the scaling coefficients (Herre et al., 1994).

Construction

MP3 files are composed of 'frames,' also known as AAU (audio access units) (Yoo et al., 2012), each with a sample length of 1152 samples. Each frame is then subdivided into two sections (or 'granules') of 576 samples each. To meet the 1152 sample requirement for a frame, some frames are zero-padded (in which zero-value samples are added) during encoding.

Each frame consists of the following parts:

- *Header*: Frame headers are 32 bytes in length and contain a synchronisation word and a description of the frame. The synchronisation word allows receivers to lock into the signal at any point in the stream, making the broadcast of MP3 files possible (Raissi, 2002). The description section contains information such as the MPEG version, layer, bit rate, and sampling frequency.
- *Side information*: This section contains the information needed to decode the Huffman-encoded data in the main data section. Huffman coding was first proposed by Huffman (1952) as a technique to minimise the number of bits used to encode data based on the frequency of occurrence of a data item.
- *Main data*: The main data consists of the scale factors, Huffman coded bits, and ancillary data. The primary function of scale factors is to reduce quantisation noise by determining the resolution of the quantiser required for each coefficient, so that the audibility of any distortion introduced during encoding is minimised (Lutzky et al., n.d.). It does this by attempting to meet or exceed the masking threshold and reduces the granularity of standard bit allocation methods. The Huffman code is decoded using the information from the previous section (Korycki, 2014).

ID3 tag

The ID3 tag is used to store text information within an MP3 file, which was initially in the form of a 128-byte tag with fields for various textual information, such as title, artist, and genre, as relates to the data present within the recording. The first version only allowed 30 characters to be stored at the end of a file, which caused issues for streaming as the entire file would have to be read before the information could be displayed. This promoted the development of an improved, dynamic version (thus allowing various amounts of data to be inserted) named ID3v2, which removed length constraints and was written to the beginning of a file to facilitate streaming (Furini and Alboresi, 2004).

Encoding process

The following is the general encoding process of an MP3 recording (Brandenburg, 1999; Raissi, 2002):

1   Frames and sub-bands
    The recording is first split into frames of 1152 samples each, before being filtered into 32 equally spaced frequency bands, dependent on the Nyquist frequency. For example, for a recording with a sample rate of 44100 Hz, the Nyquist frequency is 22050 Hz, and thus:

    22050/32 = 689 Hz

    Each spectral band will, therefore, be 689 Hz wide.
    As the bands occur in the frequency range and the filtering is applied to each frame, each band now contains 1152 samples, increasing the amount of data to 32 x 1152, so each must be decimated by a factor of 32 to return to 1152 samples.

2   FFT
    The frames are then converted to the frequency domain using 1024 and 256 point FFTs to give higher frequency resolution and information on spectral changes over time.
3   Psychoacoustic model
    The psychoacoustic model takes the data from the FFT and determines which areas of the signal are audible and which are not, thus informing the next stage in terms of the quantisation and window types required for optimal encoding.
4   MDCT
    MDCT stands for modified discrete cosine transform, and was adapted from the discrete cosine transform (DCT) used in the JPEG standard to achieve reductions in the size of digital image files.
        In order for MDCT to be performed, the frames from stage one are windowed to prevent any edge artefacts, with a window function dependent on the output of the psychoacoustic model from the previous step. MDCT is then applied to each frame, splitting each sub-band into 18, resulting in a total of 576 discrete frequency bands.

Quantisation and scaling

Quantisation and scaling are now applied to the 576 values. Quantisation is performed separately for the short and long blocks, using power-law quantisation. This form of quantisation is non-linear and quantises larger values with lower resolution.
    The scaling process then multiplies the quantised values by the scale factors.

Huffman encoding

The quantised values are then encoded using different methods, for example, run-length encoding is used to encode consecutive zero values in the upper-frequency range. Values with a magnitude below 1 use a four-dimensional Huffman code, while the remaining larger values use a two-dimensional scheme which can be split into three sub-regions, each having a separate Huffman code table. The adaptation of code tables to subregions is designed to increase coding efficiency and reduce transmission errors.

Side information

All parameters used by the encoder are then stored in the files' side information to allow the decoder to reproduce the signal.

Bit-stream formatting CRC word generation

The bit-stream is finally generated, bringing the frame header, side information, cyclic redundancy check (CRC) word and Huffman-coded data together to form a frame. Each frame represents 1152 encoded PCM samples. The frames are then compiled into a container for storage or transmission.

A CRC is a method in which data is summed into a single value which does not add any additional information to the signal, and is therefore redundant. If a check is performed and the value does not match the data, it indicates that there is an error (Stockham, 1982). As it relates to MP3 encoded files, it is used to detect errors in the header and side information (Sripada, 2006).

MPEG-2 (AAC)

Following on from the release of MP3, ISO presented AAC (Advanced Audio Codec) as part of the next generation of the MPEG standard, dubbed MPEG-2. Again, this was developed by the MPEG organisation, which at the time consisted of Dolby, Fraunhofer, AT&T, Sony, and Nokia, with the intention of delivering a standard with even better performance than MP3. It also forms the core of MPEG-4 (there is no MPEG-3). The process of encoding an ACC audio file shares similarities with the MP3 format, in that the filter bank, non-uniform quantisation, and Huffman coding stages are all included (Bos et al., 1997), but it also provides improvements through the use of coding tools or 'modules,' which can be combined to produce bitstreams at three different profiles. This allows for trade-offs in the audio quality and complexity of coding dependent on the application. The main profile is the most complex and requires all of the tools within the encoder. The LC (low complexity) profile is the primary encoder and uses some of the same tools as the main profile (so it can be decoded by a main profile decoder), but omits others. This is accepted by many as providing audio of a quality which is close to that produced using the main profile, but with significant savings in terms of memory and processing required.

The third profile is the scalable sampling rate (SSR) profile, which splits the audio into four self-contained bitstreams using frequency bands of equal width. A simple decoder can then decode 1, 2 or 3 of these bitstreams to produce a reduced bandwidth output. As with the LC profile, not all of the ACC tools are available to the SSR profile. It is generally used for low complexity applications and allows a bitstream to be decoded to PCM with a variety of sample rates.

The MPEG-2 AAC format uses a variety of coding tools to achieve a bit rate reduction, including (Watson and Buettner, 2000):

- *Filter bank tool*: The main and LC profiles use MDCT with transform length blocks switched between 1024 and 128 frequency points. SSR uses a hybrid filter bank with a polyphase quadratic filter connected in a cascade with 256 and 32 frequency points. The MDCT filter bank results in a 23 Hz frequency resolution and 2.5 ms time resolution.
- *Temporal noise shaping (TNS) tool*: This improves the temporal resolution by using frequency-domain prediction to shape noise in the time domain (Herre and Johnston, 1996). Although all profiles include the TNS tool, the bandwidth is limited in the LC and SSR profiles.
- *Joint-stereo tool*: Exploits redundancies across channels to reduce the bit rate through the M/S, intensity stereo, and coupling methods.
- *Prediction tool*: Uses backward adaptive prediction to remove signal redundancies in successive blocks. This is only available in the main profile.
- *Quantiser tool*: Quantises the spectral coefficient using a nonuniform step size of 1.5 dB, which is fixed within the scale factor band, but can vary from band to band. The common scale factor and the scale factor band determine the quantisation level for each band.

A variety of bit-rates are available within the AAC format, the maximum of which is dependent on the sample rate (including both constant and variable modes). There are, therefore, many combinations available. It is also of a higher frequency resolution than MP3, as the audio is split into 1024 bands rather than 576 (32 bands which are each further divided into 18).

The sequence for MPEG-2 AAC encoding is as follows. The audio is first sent to an MDCT filter bank, which uses block lengths of 256 and 2048 samples with a 50 per cent overlap, which equates to 23 Hz or 187 Hz bandwidths at a 48 kHz sample rate. The window function can be selected as either a Kaiser-Bessel-Derived (KBD) window or sine window as a function of the input spectrum. KBD windows were developed for use with AAC to provide improved frequency selectivity over sine windows, thus allowing a perfect reconstruction of the data (Bos et al., 1997).

The stream starts with a header, which is either ADIF (Audio Data Interchange Format) or ADTS (Audio Data Transport Stream). ADIF is generally used for file-based applications and contains one header per file. ADTS is used for serial transmission protocols and contains one header per frame to aid in synchronisation. Due to the number of improvements, only around 70 per cent of the bit rate of MP3 is required by AAC to achieve audio of a comparable quality.

MPEG-4 (AAC)

MPEG-4 AAC is designed to be a universally standardised codec, optimised for different applications and bit rates by allowing selectivity over the modules used, or employing pre-defined profiles for a specific application.

The first draft was released in 1999 and built on MPEG-2 AAC by extending the number of modules to achieve low bit rate compression. LLC is the primary general encoder, a compatible extension on MPEG-2 AAC. In addition to the MPEG-2 AAC modules, it also offers:

- *Perceptual noise substitution (PNS)*: Increases the coding efficiency by representing noise-like signals parametrically rather than coding the waveform.

- *Long-term prediction (LTP)*: Exploits time redundancy between the current and preceding frame with significantly lower complexity than the frequency domain prediction used for MPEG-2. As it is a forward predictor, it is less sensitive to bit errors within the spectral coefficients of the transmitted signal.
- *Twin VQ (transform domain weighted interleave vector quantisation)*: For good coding performance at extremely low bit rates.
- *Low-delay AAC*: Uses reduced frame lengths (512/480) instead of 1024/960 to reduce the length of the analysis window, and thus provide higher temporal resolution.
- *Error robustness*: Reduction of the perceived degradation of the decoded signal caused by bit errors. The error resistance of scale vector bandwidth large spectral coefficients are enhanced as these are the easiest to perceive.
- *Scalable audio coding*: Allows different bit rates within the same bitstream, dependent on the transmission capacity of the channel.
- *Parametric audio coding*: Uses a parametric audio coder for coding of audio signals with very low bit-rates. Frames are generated by overlapping analysis windows followed by the analysis of individual sinusoids, harmonic tones, and noise components. The perceptual model is then used to select the most redundant components (Doliwa, n.d.).

Unlike MPEG-2 AAC, an MPEG-4 AAC decoder does not need to support ADIF and ADTS. The HE (high efficiency) version is a subset of AAC and uses a combination of MPEG-2 AAC and the SBR technology used by MP3 Pro to generate high-frequency content.

Windows Media Audio (WMA)

Windows Media Audio (WMA) was developed by Microsoft, who had a long-term vision of delivering digital audio and video at a variety of bit rates for use in a variety of applications. It was first launched in 1999 as Windows Media Audio Version 1 and has since undergone several iterations (Myrberg, 2002).

There are generally three categories of WMA encoding, namely, standard, professional, and lossless. Standard is used for general-purpose compression, professional for multi-channel and high definition, and lossless for lossless compression. A separate Windows Media encoder, the Windows Media Voice Encoder, is designed for the compression of speech-only media. As new versions have been released, the specifications are subject to change, but are all designed around the same framework to ensure backwards compatibility (Microsoft Corporation, 2018).

WMA has two sample rates available for encoding, 44.1 kHz and 48 kHz (similar to AAC), at 16 bits and bit-rates between 64 and 192 kb/s. The more recent versions also support variable bit-rate encoding (Dernaika and Khavtasi, 2007). Similar to other perceptual encoding methods, MDCT is used for spectral analysis. Unlike MP3 and AAC, the window size is dependent on the sampling frequency and can take power two values between 64 and16384 (Boyars, 2016).

WMA Professional includes support for up to 24-bit, 96 kHz stereo audio, including 24–96 kbps stereo capabilities. These high specifications mean it is highly unlikely a forensic audio examiner will ever encounter a file encoded with this format.

WMA Lossless compresses at 2:1 or 3:1 ratio using a lossless compression algorithm, ensuring a bit-for-bit copy of the original audio file upon decoding.

WMA Voice is a low bit rate codec used primarily for speech recordings, although it can be used for other applications. It allows compression down to rates of 4 kbps at 8 kHz.

All WMA files use the ASF file container for storage, short for Advanced Systems Format, and designed by Microsoft for the storing and streaming of digital multimedia. The most common file extension is.wma, so named so the audio files can be distinguished from other ASF files which may also contain video (Microsoft Corporation, 2010).

The structure of the file consists of three objects, namely, a header, the data, and an optional index. The header contains the information expected of an audio file header, such as the file size and codec information. The data contains the content information stored in packets. The index is used to hold a list of index/key-frame pairs to enable applications to navigate to specific sections of the content quickly.

Vorbis

Ogg Vorbis is a perceptual audio codec created by the Xiph Foundation and released in 2002. Its development was stimulated by MP3 licensing requirements proposed in 1998 with a desire to offer a free, non-proprietary, high-quality alternative (Autti and Biström, 2004).

It uses sample rates between 8 and 192 kHz with a range of channel configurations. At 44.1 kHz, the encoder uses bit-rates between 32 kbps to 500 kbps, dependent on the quality settings and choice of bit- rate mode (constant or variable at 16–128 kbps/channel).

Similar to other perceptually encoded formats, MDCT is employed to convert the signal from the time domain to the frequency domain, followed by separation into the noise floor and residue components, quantisation, and then entropy encoding. It uses a complex encoder and a simple, low-complexity decoder (Xiph.org Foundation, 2015).

Metadata fields are referred to as 'comments' in the Vorbis format. It is similar to the ID3 standard as metadata is stored as vectors in byte strings (Dernaika and Khavtasi, 2007).

AC-3

AC-3 (Acoustic Coder 3) was developed by Dolby Laboratories and is considered a high-quality codec. High compression ratios are achieved through adaptive bit allocation to encode audio at low bit rates (Todd et al., 1994).

During encoding it first employs MDCT to transform the audio from the time to the frequency domain, before adjacent transform coefficients are grouped into sub-bands that approximate the critical bands of the human auditory system. Coefficients within each sub-band are then converted to a floating-point representation, with one or two mantissas (the non-zero part of a floating-point number) per exponent (the number of places the decimal point is moved). The exponents of this process are then encoded according to the required temporal and spectral resolution before being fed into the psychoacoustic model. The model then calculates the perceptual resolution according to the encoded exponents and perceptual parameters. The perceptual resolution and available bits are then used to quantise the mantissa. The entire process requires a close relationship between the exponent coding, psychoacoustic model, and the bit allocation (Liu and Chang, 1999).

It is one of the less common codecs to come across in digital forensic audio as it has not been widely adopted by manufacturers of portable audio devices and mobile phones.

DSS

The original codec used for DSS (Digital Speech Standard) was initially created in 1994 by Grundig and the University of Nuremberg, based on an idea from Philips. Three years later, the codec was developed into a speech standard through a collaborative effort between the Grundig, Olympus and Philips. The codec was designed to encode speech recordings using high rates of compression while maintaining quality, the result of which is files 12 times smaller than MP3 recordings of the same content. For instance, a 10-minute recording requires only 1 MB of data for storage (Grundig et al., 2005).

As it was developed by several commercial organisations, it is marketed as a Manufacturer Independent Standard. As such, it can be used by other companies, providing it is used within professional devices. There are several versions of the DSS format, including the LP (long play-back), SP (standard playback), and HQ (high quality). Sample rates of 12 kHz and 16 kHz are used, dependent on the DSS version used.

AMR

AMR stands for adaptive multi-rate and is a commercial, multi-rate audio encoder introduced in 1998, primarily used for the encoding of speech transmitted by mobile telephone. It was initially developed and standardised by the European Telecommunications Standards Institute for GSM (Global System for Mobile Communication) systems and was adopted in 1999 as the mandatory speech codec for 3GPP systems (Autti and Biström, 2004). The container format, is, therefore, based on that used for 3GP video, which is structurally based on the ISO MPEG-4 file format specification (3GPP, 2018).

3GPP (3rd Generation Partnership Project) is an international standards organisation which develops standards for mobile telephony such as 2G, 3G, 4G, and 5G. Its formation was proposed in 1998 by AT&T Wireless to develop a 3G system based on the 2G GSM system. It consists of various organisations which operate within the telephony communications industry, including British Telecom, Motorola, and Nokia (Varga et al., 2006).

AMR has eight different bit-rates, between 4.75 and 12.2 kb/s, and uses 160 samples for 20 ms frames. This results in an 8 kHz sampling rate that is filtered to a 200–3400 Hz bandwidth. The input for the encoder is taken from the User Equipment or the network of the Public Switched Telephone Network (PSTN) via an 8-bit a-law or u-law to 13-bit uniform PCM conversion (3GPP, 2002). It does not support encoding of multi-channel audio content into a single bit-stream, but can be used to encode and decode individual channels separately.

As its primary medium is transfer over telephony communication (including some VOIP applications), its priority is the reduction of bandwidth while ensuring intelligibility of speech. For this to be achieved, it removes areas of redundancy which other formats not explicitly designed for speech are unable to tolerate. It does so through various technologies used to reduce the signal bandwidth during quiet periods. As the person talking alternates during a telephone conversation, a technique called Source Controlled Rate encodes regions of the signal which are determined by a voice activity detection (VAD) technique to contain only background noise at lower bit rates than for normal speech. VAD uses a set of parameters computed by the AMR speech encoder which enables it to determine whether each 20ms frame contains speech. Another technique is termed comfort noise generation, in which artificial noise is synthesised in regions determined to contain no speech by the VAD. As transmissions can contain errors, and packets can be dropped, an error concealment mechanism is also implemented to mask the impact of lost frames. When frames are dropped, the speech decoder is informed, and error concealment is initiated using a set of predicted parameters for speech synthesis. If there are several lost frames, a muting technique is used to indicate that the transmission has been interrupted. It also lowers the bit-rate as interference increases, thus allowing more error correction to be applied (Bessette et al., 2002).

Aside from its use in mobile telephone calls, a file format for spoken audio also exists, leading to mobile phones adopting the technology to store and share speech recordings. It is used by some free software applications and is also used for sending voice messages through SMS.

Perceptual encoding artefacts

As the process of perceptual encoding removes data required for an accurate reconstruction of the analogue signal, there is an inevitable introduction of artefacts associated with this data loss to those with a critical ear. A forensic audio examiner needs to be able to identify and understand these artefacts to aid in determining whether they are consistent with editing, or can be explained by the perceptual encoding processes which the audio has undergone. It is not uncommon for work to be instructed based on an artefact heard by a layman who believes it to be an edit point. Somewhat rarer, although still relatively common, is for somebody to interpret an artefact as distorted speech.

Perceptual encoding artefacts vary based on factors such as the bit rate, sample-rate, bit rate mode, and audio content. In a general sense, the lower the specifications used, the more artefacts that will be introduced as more data is removed to meet these specifications. They can be periodic or seemingly random, and due to the filter bank are not only confined to the regions which contain audio content. The result can be the introduction of non-harmonic distortions and noise in specific frequency ranges (AES Technical Committee of Coding of Audio Signals, 2001). Although many of the artefacts are audible, others can only be perceived through visual analysis of the spectrogram, while some can be both visually and audibly detected. The following presents an overview of the most common artefacts of perceptual encoding.

Bandwidth reduction

For signals containing content across frequencies which are beyond the capabilities of the encoder, some frequency bands which exist in the original may not be represented in the encoded version. This most commonly occurs at high frequencies, and if the loss is not consistent over several frames, it is more audible than when repeated for a prolonged period due to the rapid and stark transitions between frames. These areas of data loss are known as spectral valleys (Brandenburg, 1999).

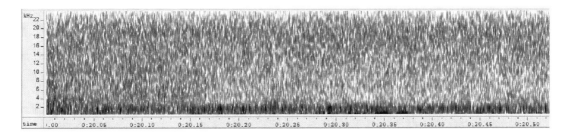

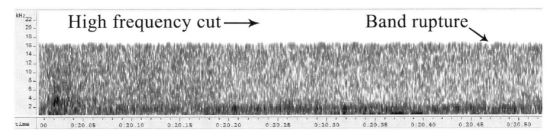

Figure 6.4 High-frequency cut. Upper: Original uncompressed signal, Lower: MP3 192 kbps encoded version of the same signal

Source: Captured from Wavesurfer.

It is also common (but not ubiquitous) for codecs to implement a bandwidth reduction to 16 kHz or below, known as spectral clipping (Figure 6.4). The compression algorithm saves bits by quantising higher frequency bands with zero bits and then using these bits in the lower frequency range to which our ears are more sensitive. This is easily achieved as the frequencies have already been split into separate bands through the use of the filter bank. It is also used to reduce the risk of birdies (which will be covered in its own section) due to the aforementioned spectral valleys. The presence of artefacts related to spectral clipping can be audible as a muffled character to the sound.

Pre-echoes

When audio is decoded, a synthesis filter bank reconstructs the audio for each frequency line or band. As quantisation noise is introduced to each original frequency line, and the length of this noise is equal to the length of the synthesis window, the noise is spread across the range of the window. For stationary signals, this noise is hidden within the signal thanks to a pre-determined masking threshold, but when there is a transient present within the length of the window, the noise is unmasked in the silence preceding the attack. This effect causes a short hissing noise and a blurring of the attack portion of the signal (Iwai, 1994).

Double-speak

At low bit rates and low sample rates, a mismatch between the time resolution of the encoder and the original signal results in a single voice that can appear as if it has been recorded twice and overlaid. AAC introduced the process of temporal noise shaping (TNS) to enhance the time resolution of the filter bank, thus reducing this effect (Brandenburg, 1999).

Birdies

When frequency lines containing little information and are quantised to zero values, the resultant dips in the audio spectrum are known as spectral valleys. They are caused by the unsuitable bit allocation or excessive masking energy estimates from the psychoacoustic model, and as perceptually encoded audio assigns bits to each block dependent on the information contained within, spectral coefficients can appear and reappear based on the bit demand. The resultant audible artefacts are a change in timbre and the presence of high-frequency energy variations between frames known as 'birdies' (Erne, 2001). The disparity between different scale factors being applied to consecutive frequency bands can also result in distinctive ruptures between the bands when viewed within a spectrogram.

Speech reverberation

Due to limitations with the Fast Fourier Transform algorithm, the trade-off for a high resolution in the time domain is a reduced resolution in the time domain and vice versa. Although this is not a problem for stationary signals, such as extended tones, as isolated peaks can be quantised separately to ensure the spectral peak remains lower than the masking threshold within each band, it is an issue for non-stationary signals.

The types of signals that occupy a broad spectrum also require a high temporal resolution due to their constant transient nature. The result is 'speech reverberation' artefacts, which become more audible for speech signals in which the coding algorithm uses large transformation lengths (Erne, 2001).

Loss of stereo image

Perceptual encoding uses methods to take advantage of binaural psychoacoustic effects, the most common of which are mid/side (M/S) and intensity stereo coding.

As the spatial perception of frequencies above approximately 4 kHz relies on the analysis of energy-time envelopes rather than the waveform of a signal, intensity stereo encodes the envelope rather than the waveform. This is achieved by encoding a common set of spectral coefficients in a 'carrier signal' shared between several audio channels instead of each having their own channel. When decoded, the carrier signal is scaled for each signal channel to match the original average envelope for each coder frame based on scaling information stored within the file. As they are all scale versions, they have the same envelope structure for the duration of a block. For a transient signal with a disparity of envelopes in different channels, this causes an issue as the original distribution of the envelope onsets between the coded channels cannot be recovered (Erne, 2001).

Phase shift

As perceptually encoded recordings take advantage of perceptual masking, our sense of the amplitude of the remaining frequency content can change as some sounds become attenuated or

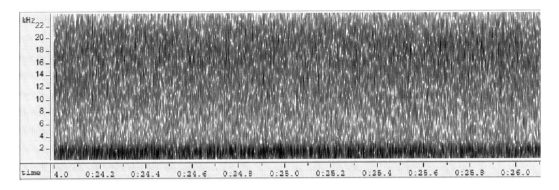

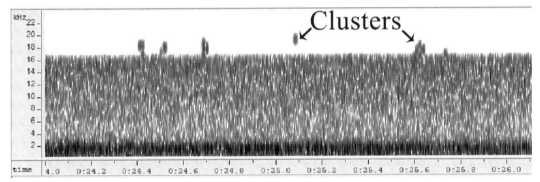

Figure 6.5 Cluster artefacts. Upper: Original uncompressed signal, Lower: MP3 192 kbps encoded version of the same signal

Source: Captured from Wavesurfer.

compressed. This can result in some sounds appearing louder than intended due to their relativity to attenuated sounds, both temporally and spectrally. When frequency content is smeared over time, its average level increases, resulting in perceived changes to the timbre and frequency masking of the audio (Corbett, 2012).

Holes and clusters

Time/frequency areas which are deemed by the perceptual model to contain little information are quantised at low bit-rates, which, when occurring over large regions, can become visible in a spectrogram as holes among otherwise seemingly content-rich areas (Figure 6.5). The visibility of these 'clusters' is due to the bit-rate disparity in adjacent regions (Hennequin et al., 2017).

Summary

Audio data can exist in several formats, from audio-only files to a single stream within a video. There are also various encoders available, and a variety of containers for the encoded audio to be stored. From a forensic standpoint, the publication of standards and specifications relating to these formats can be used to our advantage as the method used can be understood and exposed to reveal traces of manipulation, from local edits through to conversion from other formats. It can also aid those involved in the capture of audio to ensure the optimal specifications are selected for the task at hand.

References

3GPP, 2002. CWTS-STD-DS-26.071 (Release 5).

3GPP, 2018. Transparent end-to-end packet switched streaming service 26.244.

AES Technical Committee of Coding of Audio Signals, 2001. Perceptual audio coders: What to listen for. Audio Engineering Society, New York.

Apple Computer, Inc, 1989. Audio Interchange File Format: 'AIFF' (No. Version 1.3). Apple.

Apple Computer, Inc, 2016. Quicktime file format specification. Apple.

Autti, H. and Biström, J., 2004. Mobile audio – from MP3 to AAC and further. HUT, Telecommunications Software and Multimedia Laboratory, Helsinki.

Aziz, P.M., Sorensen, H.V., and von der Spiegel, J., 1996. An overview of sigma-delta converters. IEEE Signal Processing Magazine 13, 61–84.

Bessette, B., Salami, R., Lefebvre, R., Jelinek, M., Rotola-Pukkila, J., et al. 2002. The adaptive multirate wideband speech codec (AMR-WB). IEEE Transactions on Speech and Audio Processing 10, 620–636.

Bos, M., Brandenburg, K., Quackenbush, S. Fielder, L. Akagiri, K., et al., 1997. ISO/IEC MPEG-2 Advanced Audio Coding. Journal of the Audio Engineering Society 45, 789–814.

Boyarov, A. 2016. Detection and analysis of tracks compressed audio signal coded MP3, AAC, WMA and Vorbis. Theory and Practice of Forensics 1 (41).

Brandenburg, K., 1999. MP3 and AAC explained. Paper presented at 17th International Conference: High-Quality Audio Coding, Audio Engineering Society.

Corbett, I., 2012. What data compression does to your music. Sound on Sound. Available at: www.soundon-sound.com/sound-advice/what-data-compression-does-your-music-media

Cummiskey, P., Jayant, N.S., and Flanagan, J.L., 1973. Adaptive quantization in differential PCM coding of speech. Bell System Technical Journal 52, 1105–1118.

Deatherage, B.H. and Evans, T.R., 1969. Binaural masking: Backward, forward, and simultaneous effects. The Journal of the Acoustical Society of America 46, 362–371.

Dernaika, G. and Khavtasi, S., 2007. Multimedia codec evaluation and overview. Blekinge Institute of Technology, School of Engineering, Blekinge, Sweden.

Dimino, G. and Parladori, G., 1995. Entropy reduction in high quality audio encoding. Paper presented at the 99th AES Convention, Audio Engineering Society, New York.

Doliwa, P., n.d. MPEG-4 Advanced Audio Coding 13. Available at: pdfs.semanticscholar.org/6f87/dce0d-7912f6d5f7026f7131013fd22d34e4d.pdf

Erne, M., 2001. Perceptual audio coders: 'What to listen for' Paper presented at the 111th AES Convention, Audio Engineering Society, New York.

FFmpeg, n.d. FFmpeg documentation: Codecs list. Available at: ffmpeg.org/ffmpeg-codecs.html

Fletcher, H., 1940. Auditory patterns. Review of Modern Physics 12, 47–65.

Furini, M. and Alboresi, L., 2004. Audio-text synchronization inside MP3 files: A new approach and its implementation. Paper presented at the 2004 1st IEEE Consumer Communications and Networking Conference, IEEE, Las Vegas,

Gadegast, F., 1996. The MPEG-FAQ (No. 4.1). PHADE Software, Germany.

Gibbs, M., 1999. Getting a handle on RIFF audio and video formats. Network World. Available at: edition.cnn.com/TECH/computing/9908/06/riff.idg/index.html

Grundig, Olympus and Philips 2005. .dss (Digital Speech Standard) = MP3 for Speech. Press release.

Hacker, S., 2000. MP3: The definitive guide. O'Reilly Media, Sebastapol, CA.

Hennequin, R., Royo-Letelier, J., and Moussallam, M., 2017. Codec independent lossy audio compression detection. Paper presented at the 2017 IEEE International Conference on Acoustics, Speech and Signal Processing (ICASSP), IEEE, New Orleans, LA.

Herre, J., Brandenburg, K., and Lederer, D., 1994. Intensity stereo encoding. Paper presented at the 96th AES Convention, Audio Engineering Society, Amsterdam, The Netherlands.

Herre, J. and Eberlein, E., 1992. Combined stereo coding. Paper presented at the 93rd AES Convention, Audio Engineering Society, San Francisco.

Herre, J. and Johnston, J.D., 1996. Enhancing the performance of perceptual audio coders by using temporal noise shaping (TNS). Paper presented at the 101st AES Convention, Audio Engineering Society, Los Angeles.

Huffman, D.A., 1952. A method for the construction of minimum-redundancy codes. Proceedings of the IRE 40, 1098–1101.

Iwai, K.K., 1994. Pre-echo detection & reduction. MIT Press, Cambridge, MA.

Jayant, N.S., 1974. Digital coding of speech waveforms: PCM, DPCM, and DM quantizers. Proceedings of the IEEE 62, 611–632.

Kaur, A. and Dutta, M.K., 2018. An optimized high payload audio watermarking algorithm based on LU-factorization. Multimedia Systems 24, 341–353.

Koenig, B.E. and Lacey, D.S., 2014. Forensic authenticity analyses of the metadata in re-encoded WAV files. Paper presented at the 54th International Conference: Audio Forensics, Audio Engineering Society, London.

Korycki, R., 2014. Authenticity examination of compressed audio recordings using detection of multiple compression and encoders' identification. Forensic Science International 238, 33–46.

Liu, C.-M. and Chang, W.-W., 1999. Audio coding standards. Available at: www.mp3-tech.org/programmer/docs/AudioCoding.pdf

Lutzky, M., Schuller, G., Gayer, M., Krämer, U., and Wabnik, S., n.d. A guideline to audio codec delay. Paper presented at the 116th AES Convention, Audio Engineering Society, Berlin, Germany.

Microsoft Corporation, 2010. Advanced Systems Format (ASF) specification (No. Revision 01.20.05). Available at: docs.microsoft.com/en-us/windows/win32/wmformat/advanced-systems-format--asf.

Microsoft Corporation, 2018. Choosing an audio codec. Available at: docs.microsoft.com/en-us/windows/win32/medfound/choosinganaudiocodec

Microsoft Corporation and IBM Corporation, 1991. Multimedia programming interface and data specifications 1.0. Available at: www.tactilemedia.com/info/MCI_Control_Info.html

Musmann, H.G., 2006. Genesis of the MP3 audio coding standard. IEEE Transactions on Consumer Electronics 52, 1043–1049.

Myrberg, M., 2002. Windows media in consumer electronics: Past, present and future. Paper presented at the 17th UK Conference: Audio Delivery, Audio Engineering Society, London.

Pohlmann, K., 2005. Principles of digital audio, 5th edn. McGraw-Hill, New York.

Pope, S.T. and van Rossum, G., 1995. Machine tongues XVIII: A child's garden of sound file formats. Computer Music Journal 19.

Raissi, R., 2002. The theory behind MP3. Available at: www.mp3-tech.org/programmer/docs/mp3_theory.pdf

Reeves, A.H., 1965. The past, present and future of PCM. IEEE Spectrum May, 58–63.

Roussev, V. and Quates, C., 2013. File fragment encoding classification: An empirical approach. Digital Investigation 10, S69–S77.

Sripada, P., 2006. MP3 decoder in theory and practice. Blekinge Institute of Technology, Department of Signal Processing and Telecommunications, Blekinge, Sweden.

Stockham Jr, T.G., 1982. The promise of digital audio. Paper presented at the 1st International Conference, Audio Engineering Society, New York.

SWGDE, 2017. Technical notes on FFmpeg (No. 1.0). Available at: www.swgde.org/documents/Released For Public Comment/SWGDE Technical Notes on.

Texas Instruments, 2002. G.726 adaptive differential pulse code modulation (ADPCM) on the TMS320C54x DSP. Available at: www.ti.com/lit/an/spra118/spra118.pd

Todd, C.C., Davidson, G.A., and Davis, M.F., 1994. AC-3: Flexible perceptual coding for audio transmission and storage. Paper presented at the 96th AES Convention, Audio Engineering Society, Amsterdam, The Netherlands.

Varga, I., De Lacovo, R.D., and Usai, P., 2006. Standardization of the AMR wideband speech codec in 3GPP and ITU-T. IEEE Communications Magazine 44, 66–73.

Watson, M.A. and Buettner, P., 2000. Design and implementation of AAC decoders. IEEE Transactions on Consumer Electronics 46, 819–824.

Xiph.org Foundation, 2015. Vorbis I specification. Available at: xiph.org/vorbis/doc/Vorbis_I_spec.html

Yoo, B., Park, J., Lim, S., Bang, J., and Lee, S., 2012. A study on multimedia file carving method. Multimedia Tools and Applications 61, 243–261.

Ziegler, T., Ehret, A., Ekstrand, P., and Lutzky, M., 2002. Enhancing MP3 with SBR: Features and capabilities of the new MP3PRO algorithm. Paper presented at the 112th AES Convention, Audio Engineering Society, Munich, Germany.

Zwicker, E., 1961. Subdivision of the audible frequency range into critical bands (Frequenzgruppen). The Journal of the Acoustical Society of America 33, 248.

Chapter 7

Preparatory analysis

Introduction

Before any forensic audio examination can commence, there are several preparatory steps which must first be performed (Reith et al., 2002). As much care and consideration should be given to these stages as any other part of an examination as any errors here could affect all future analyses or processes which follow, and subsequently, the reliability of the final outcome.

Exhibit imaging

Although in some respects digital evidence is less volatile than some of the more traditional forensic sciences, such as fingerprints and DNA (which can become contaminated solely through physical contact), it is still an extremely fragile format and can be altered beyond repair relatively easily (National Institute of Justice, Office of Justice Programs, 2004). For this reason, original exhibits should be treated with extreme caution, and this cannot be emphasised enough. Errors such as the pressing of the wrong button on a recorder, placing the evidence within a magnetic field, or using an incompatible power adapter are simple to make but can be irreversible. In a worst-case scenario, these types of mistakes could result in data loss, and potentially (depending on the evidence and its bearing on a case), the wrongful conviction or release of an individual.

Exhibit imaging pertains to the creation of a bit-stream digital copy of an exhibit. The bit-stream refers to the binary digits of which the exhibit is composed, ensuring an accurate and complete replica of the original version is created. The procedure for imaging of evidence is highly dependent on the format and the manner in which the recording was received. Extreme caution should always be practised with evidence provided on the original device with which it was purported to have been captured, but there is no harm in treating every exhibit received as the 'original' version, regardless of whether it is a digital copy or not. In treating each exhibit in the same manner, it ensures consistency, good practice, and mitigates against any issues concerning the chain of custody (Casey et al., 2009). The formats in which digital audio evidence can be received will be in one of two media, that of physical or digital.

Physical

This relates solely to original audio evidence contained on recording devices. There are several possibilities for imaging in these cases as the recording may exist on a device with USB connectivity, removable external memory, both, or neither of these possibilities. As specific considerations are required for mobile phones and computers, they are outside the scope of this book, and further information should be sought from other sources regarding how they should be processed. Recordings which have been previously extracted by a third party will generally have been imaged to a storage format such as a disc or an external USB device (for instance, a USB flash drive or external hard drive).

Upon receiving an original device on which the audio is purported to have been captured, the following steps must be taken to ensure the integrity of the exhibit is not compromised (SWGDE, 2016):

1   Information should first be obtained from the instructing party as to whether any other forensic work is to be instructed, such as the recovery of fingerprints. This allows precautions to be taken to ensure the physical evidence is not contaminated. All latent forensics should have been performed before the device is provided for a digital forensic examination, but best practice would always be to ask the question.
2   Original recording devices should be photographed upon receipt and any external damage to the unit (which should be ensured is visible in the photographs) documented.
3   The user manual should be thoroughly reviewed to gather an overview of the functioning of the device to ensure buttons which could affect the recordings are not pushed, that the correct power adaptor is used, and the optimal method for the extraction of the data is used.
4   Attempts should be made to turn the device on. If the power has drained due to the time elapsed since it was first seized, the relevant power lead must be sought. Steps must always be taken to ensure power adaptors are of the correct voltage and current so as not to damage the circuitry of the device.
5   A case folder must be created on the workstation pertinent to the case, with a structure which includes an exhibit folder to store the digital working copy.
6   A write blocker should always be connected to the workstation and verified to be working by attaching a test device such as an external hard drive and attempting to write to the device, change file names, and image a file from the device to the workstation.
7   If the workstation does not have a regularly updated virus scanner always running as a background process, one should be activated before connecting the exhibit to ensure the examiner is alerted to any viruses which may be contained on the exhibit.
8   Once the write blocker is shown to be working correctly, the exhibit can be connected to the workstation and files imaged from the device to the exhibit folder. It is good practice to image all recordings, thus ensuring no data is missed which may be relevant to the work instructed, and to prevent imaging of the device a second time if additional work is instructed in the future.
9   The file options of all exhibits imaged to the workstation should be changed to read-only, thus removing any write permissions. This prevents accidental changes being made to the data.
10 A log of the exhibit should be created, documenting information such as the evidence bag number, the client name, the case name, the date the exhibit was received, the date imaged, and the name of the examiner who performed the imaging.
11 The hash checksums of both the original evidence and that imaged to the workstation should be compared to ensure exact bit-stream copies have been made. When a write blocker is in

use, this will not affect the data on the device as the hash checksum is a calculation based on the file, so does not change the actual data. The write blocker also prevents any changes to the access date, ensuring the original evidence metadata is preserved.

12 Once the imaging has been completed, the recording device can be disconnected, powered down, and placed within a labelled evidence bag.

13 The exhibit bag must be sealed and placed in secure storage such as a safe, and kept away from any electromagnetic radiation and other potential risks to its integrity.

14 It is best to keep the exhibit for the duration of the examination in case further information is required from the device. Once the work is complete, the exhibit should be returned to the party who provided it and the date it was returned internally logged.

The above assumes a portable digital recorder, but the process can be applied to any form of external media. Mobile phones are increasingly more complicated due to automated backups and the requirement for passcodes. A mobile phone examiner should, therefore, be recruited to perform an extraction of audio files from these types of devices to avoid any issues relating to contamination and to ensure the chain of custody is maintained. The pertinent files can then be provided to an audio examiner, whether they exist as audio, video, or voicemail recordings.

Digital

As we are concerned solely with digital audio, the distinction should be made as to the difference between physical and digital data. The previous subsection refers to digital data provided via a physical medium. This section relates to recordings provided via digital means such as from the cloud or an email attachment. As digital methods create a digital copy of the file automatically when copied to a workstation, they are, by definition, bit-stream copies, but best practice would be to treat them as the original. If the party who provided the exhibit deletes their copy, and the device from which they were obtained is damaged, having a copy on the workstation in the original state in which it was provided will ensure that a copy of the original version is still available. The same general process should be applied as for physically received exhibits, namely, imaging the files to an exhibit folder, implementing write protection, and making a working copy if processing is to be performed.

A point to note about exhibits is that the most original version should always be requested to ensure the most robust possible chain of custody. If a recording is provided via email, but the provider states that the device is available, the evidence should be extracted directly from the device rather than using the version supplied via email to ensure all possible steps have been taken to work with the most original version. Although digital data does not degrade when copied (unlike electrical or magnetic recordings which degrade with each iteration), by extracting the audio from the purported capture device the examiner can be confident that the audio is, in fact, the original version (although an authentication triage should be carried out regardless).

In sourcing the most original exhibit, or at least a bit-stream copy of it, there are limited opportunities for manipulation of the recording which may have occurred since the capture. If this is not the case, the possibility always exists that the original recording was copied from a device to a computer by a party with an interest in the case, who then purposely manipulated it before providing it as evidence. Conversion of an uncompressed WAV PCM version to MP3 is also an all too common occurrence as many people believe that is the most ubiquitous format and thus will play on all computers, but do not understand the quality degradation this may cause to the content.

A prime example of this relates to an enhancement case in which I received an audio recording captured using a mobile phone, which had subsequently been extracted and 'enhanced.' When I requested a copy of the original, I was repeatedly sent the enhanced version, which was severely clipped. I assumed this was because the original was clipped and the examiner who performed the enhancement had decided not to de-clip the recording for some reason. When I finally received a bit-stream working copy of the original, it was not clipped in the slightest. It was, in fact, of a much higher quality than the 'enhancement.'

In many cases, access to the original recording device is not a viable option, for instance, when access to specific systems is not possible, such as emergency call centre recording software and that used for undercover recordings. Access can also be a problem when the device belongs to a private citizen or has since been lost or destroyed. In these cases, the steps taken to attempt to obtain the original versions, the response received, and any limitations related to the version provided should be documented to be fully transparent. An example limitation would be that it may be possible for another examiner using the original version to produce an enhancement of higher quality.

Instruction

Instructions are provided by the party for whom the work is being undertaken, and can vary, based on the type of case and the reason the work is required. For complete clarity and conciseness, it is recommended that the instruction is broken down, for example, imagine the following is received from the instructing party:

> We have a recording of an alleged bribery attempt, but it is low in volume and was recorded in a room with a noisy air conditioning system. It is, therefore, difficult to hear what is being said. We also believe that there is a section which has been removed from the recording based on one of the parties' recollection of the meeting.

The points of instruction can, therefore, be refined as:

1 authentication of a single (1) audio recording;
2 enhancement of a single (1) audio recording.

These points can then be relayed to the instructing party to ensure their request has been understood. It should be made clear that an instruction should be used as guidance as to the requirements of the instructing party. The work process should be completely impartial and unbiased, regardless of who the instructing party is and the conclusions they hope will be reached.

Summary

To ensure a forensic audio examination is performed with the highest integrity possible, and in accordance with best practice guidelines, preparation is critical. It could be argued that this stage is the most important, as it is here that mistakes such as changes to the original exhibit cannot be rectified. The preparatory stage also has an influence on the rest of the examination so it must be undertaken logically. Only once working copies have been made, the original evidence secured, and the request made by the instructing party fully understood and agreed, should the examination in chief begin.

References

Casey, E., Ferraro, M., and Nguyen, L., 2009. Investigation delayed is justice denied: Proposals for expediting forensic examinations of digital evidence. Journal of Forensic Sciences 54, 1353–1364.

National Institute of Justice, Office of Justice Programs. Forensic examination of digital evidence: A guide for law enforcement: (378092004-001). U.S. Department of Justice, Washington, DC.

Reith, M., Carr, C., and Gunsch, G., 2002. An examination of digital forensic models. International Journal of Digital Evidence 1(3), 1–12.

SWGDE, 2016. Best practices for forensic audio. October 8, 1–28. Available at: www.swgde.org/documents/published

Chapter 8

Audio acoustics

Introduction

Acoustics is the science of physics relating to sound, and as would be expected, there are several areas in which it can be applied to forensics. As this book primarily pertains to the digital aspects of audio forensics, only a brief overview of the primary acoustic topics and their application to audio forensics will be provided. The perception of sound is not within the immediate scope of the work performed by a digital forensic audio examiner (for example, opining on whether an individual is likely to have heard a specific sound or not), so will not be considered in any form.

The sound made by a firearm (and the analysis of such) form a perfect example of an application of acoustics to forensics as it is suitable for a wide array of acoustic analyses and also exposes their associated limitations. The focus of this chapter will, therefore, be the application of acoustic forensics to gunshot analysis, but it should be considered that the techniques used can be applied to a plethora of other events which occur in the acoustic environment. Gunshots provide a perfect example in terms of their acoustic characteristics, and there is a relatively large pool of research into the topic. The analysis of gunshots is also extremely common in some parts of the world, in particular, the USA. Even though digital recordings are a representation of the acoustic environment in which they were captured, and although other types of examinations such as enhancement and authentication take acoustic factors into account, it is not the primary focus of the examination.

Before venturing further into acoustics, a crucial point must first be understood to prevent the misinterpretation of an acoustic signal. A recording of an acoustic sound source is not only a representation of the sound source and the environment in which it was captured. The device used to make the recording is also hugely influential on the characteristics of the captured sound (Beck et al., 2011). Factors relating to the capture device include the microphone, the sampling rate, the bit-depth channel configuration, and the encoding algorithm used. As the majority of consumer-level devices are designed for the capture of speech, they are not optimised for other signals, and especially not for gunshots, which are both extremely transient in nature and high in magnitude. Environmental factors include reflections from the ground and the surrounding buildings, the signal to background noise ratio, microphone positioning (including whether it is static or moving), and any physical shielding of the microphone.

As there are principally two areas which impact the captured signal, an examiner requires expertise in both acoustics and digital signal processing in order to be competent in audio acoustics.

Having this knowledge gives examiners the ability to separate artefacts caused by the device, and thus allow the focus of the analysis to be on the acoustics and the environment. If an examiner does not thoroughly understand how each can affect the signal, it may be that artefacts caused by the capture device are misinterpreted as being caused by the acoustic environment and vice versa, potentially leading to erroneous conclusions. The effects of the device and the environment can also cause a significant transformation of the sound and result in a loss of information in both the time and frequency domains. This can lead to a reduction in the number of analyses that are possible due to the suitability of a recording for certain types of analysis. For example, it may be possible to provide an opinion on the number of shots fired, but not the location of the firearm.

The specific use of acoustic analysis of gunshots has its roots in the military, in which the tri-angulation of enemy gunfire was first used. One of the earliest techniques of this kind was known as Sound-Ranging, developed by William Lawrence Bragg, a British military officer and physicist. The system was designed to perform calculations based on the differences between signals cap-tured from microphone arrays, and by the time the Second World War had begun, most countries were using some form of Sound-Ranging method (Van der Kloot, 2005). Nowadays recordings used for these types of analysis are captured from mobile phones, emergency call centres, or dash-mounted and body-worn cameras to inform opinions in relation to a number of questions, such as: 'Who fired first? How many shots were fired? What calibre of weapon was used?'

In the field of acoustics, there is often a requirement for corroborative information relating to the event, something that would normally not be requested for the majority of other examina-tions, in a bid to mitigate against biases. Requests generally include being provided with any video imagery (to aid in providing information with regards to the position of the source of a sound), the locations and nature of the microphones (for example, stationary or moving), and the number of people involved (to limit the number of potential sound sources contained within a recording). There is a fine line between corroborative and biasing information, so great care should be taken to request only that which is necessary for the work being performed. When and how any of this information was relied upon should also be documented within both working notes and the final report.

Overview of sub-disciplines

There are several distinct areas of acoustic analysis, and each will be considered in isolation in the following sub-sections.

Timing information

The timing at which specific acoustic events occurred can be critical to some cases, especially those where there is a dispute over the chronology. It can also be used to aid in determining the type of firearm used by measuring the interval between consecutive shots and providing an opin-ion on who fired first (Beck, 2019a; Maher, 2016). For example, imagine a case where a single gun has been recovered from the scene, but the defendant claims he was not acting alone and had an accomplice who also fired. As the weapon is known, and as timing information relating to each firearm is different, further research into the specific gun used is required to measure the shortest possible interval between shots, by having a firearms expert fire off a series of consecutive shots as quickly as they possibly can (Beck, 2017). Measurements can then be made to determine the shortest possible interval between the shots and comparisons made against the periods between shots on the evidence recording to determine if the defendant's claims are correct. If it is found that the weapon can fire, say, one round every 200 ms, and the gap between audible gunshots is 100 ms, this would be consistent with the information provided by the defendant. Obviously, there may be other considerations, for example, 'Did he have a weapon in each hand?,' but other types of analysis could then be used to determine whether each shot was fired from the same location.

Number of acoustic events

Analysis to determine the number of acoustic events which occur usually applies to a sound of the same acoustic character, and is used to answer questions such as 'How many times was the weapon fired?' This type of analysis first requires the detection of potential gunshots sounds within the recording before further analysis can be performed to determine whether the sounds are in fact gunshots and the number of times that they occur.

Identification of type of sound

This type of analysis is used to determine whether a sound is consistent with being from a general sound source (for example, a gun), rather than a specific sound source (for example, a particular make and model of a handgun). Although it can be relatively clear-cut in situations where there is a good signal to background noise ratio, and the desired signal is of a strong magnitude, it can also be extremely challenging. If similar sounds are occurring within the same period, and the desired signal occurs at a distance, while being out of the direct line of the microphone, convolution can cause the captured sound to become too complex to draw a reliable conclusion.

Identification of the source of the sound

As demonstrated in the previous subsection, it can be challenging to reliably determine the type of sound captured, never mind the specific source, but in who fired first scenarios, this is the type of analysis which would be sought. The acoustic analysis of gunshots is an area which, in an ideal world, could provide an answer as to the exact make and model of gun, or even the calibre of the weapon used, but in reality, this is not an easy task. The biggest problem during casework is caused by the mismatch between the types of signal being captured and the recording devices used. As described earlier, audio is a combination of the acoustic environment and the recording device, and when devices are pushed to their limits (as they are for high magnitude, extremely transient sources, such as gunshots), the way in which they react is hard to predict. It can, therefore, be very difficult to determine if specific artefacts are related to the type of gun, the capture environment, the capture device, or a combination of all three.

The first problem is in the time domain. As the rise time of a gunshot is only a few milliseconds (Maher, 2007), an accurate representation of the sound source is not possible using consumer-grade devices. There is, therefore, very little data captured which would provide characteristics specific to a single type of gun. The second problem is with regards to the magnitude, as the deafening sound of a gunshot causes clipping when the capture device is within close proximity of such. Not only will much of the information pertaining to the sound emitted go uncaptured, but the clipping will also cause distortion in the process. Further to this, gunshots are directional and do not propagate sound spherically as may be expected, with many having a broadband difference of -20 dB SPL behind the barrel compared to in front and on-axis. Research has also found that the difference in the level and waveform of an on-axis and off-axis capture of the same firearm is significantly higher than two different guns at the same azimuth (Maher and Routh, 2015). Other research by the same authors found that repeated shots from the same firearm, in the same position show variability. This is surmised to be due to the acute differences between each bullet, the temperature of the gun barrel (which will increase the more shots are fired as the explosion of gunpowder heats the weapon), and the slight variation in the position of the weapon held by the shooter (Maher, 2019).

Despite the various difficulties, it should be noted that when there are a limited number of weapons, which are identifiable, and a significant disparity between the sounds emitted by each, there is an increased possibility of a conclusion being rendered. For example, if a police officer had a small calibre handgun and a suspect had a shotgun, the differences would be much greater than if both had a handgun.

Location of the sound source

For the location of a sound source to be determined, known characteristics are used based on reverberation, sound quality, and information from multiple channels. For moving sources, the analysis can be somewhat tricky in comparison to a fixed-point source as the orientation of the firearm can affect the sound level and, thus, the characteristics of the capture signal (Maher, 2010). As the speed of sound is relatively constant, time of arrival differences can be interrupted as differences in linear distance to aid in providing information as to the spatial location of the source.

The most commonly known method is multilateration (Maher and Hoerr, 2018). This is based on the theory that when two or more microphones are used, there is a difference in the distance between the sound source and each receiver (Figure 8.1). This allows the time difference of arrival (TDOA) to be used to identify the possible sound source locations, providing the location of the microphones is known. The higher the number of microphones, the more accurate and reliable the analysis. This is performed in three stages, as follows:

1   Estimate position of microphones
    The microphone positions are required in order to make a multilateration estimate, which can be obtained using video, diagrams, or other sources that have been provided which relate to the captured audio. It is essential to identify uncertainties, and if necessary, remove them from the analysis at this stage to prevent errors in the results.

2   Synchronisation
    Information in relation to the formats of the recordings must be documented, and if there are compressed recordings within the dataset, these must be decoded into a PCM format to

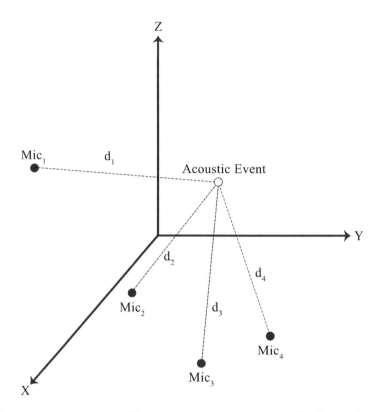

Figure 8.1 Multilateration theory, in which 'd' represents the distance travelled by the signal to each microphone

mitigate against the effects of compression. The recordings can then be transcoded to ensure they are all of the same sample rate. A synchronisation point must then be determined. The most common point is a signal transmitted to police radios, as all individuals wearing one will receive the signal at the same time without any delay caused by the TDOA.

3 Time difference of arrival calculation

For the TDOA to be estimated, the corresponding time instants must be determined within the audio waveforms. Some waveforms will not have represented the signal in a manner suitable for this to be performed so they cannot be used, and others may not be in a direct line of the source, so would be reflected and diffracted. As all captures will influence the results, great care must be taken throughout to ensure only recordings suitable for this type of analysis are used. The multilateration estimate can then be calculated.

The acoustics of a gunshot

When a gunshot signal is represented within the time domain, three distinct regions are visible, each corresponding to an acoustic event.

Muzzle blast

Firearms use hot expanding gas caused by the combustion of gunpowder to expel a bullet out of the barrel. As the gas escapes from the end of the barrel, it causes a disturbance to the surrounding acoustic pressure, resulting in what is referred to as a 'muzzle blast.' This is typically less than 3 milliseconds in duration and propagates in all directions, although the magnitude is strongest in the direction the barrel is pointed. The signal can exceed 150 dB in the immediate vicinity of the blast, but when fitted with a silencer or sound suppressor, the magnitude is significantly reduced (Freytag et al., 2005; Koenig et al., 1998). The acoustical characteristics of the muzzle blast depend on various factors, including the type and size of the weapon, the ammunition used, the direction of the barrel axis in relation to the microphone, the presence of reflections, and diffractions from obstacles (Beck, 2019b).

Shockwave

If the bullet is travelling faster than the speed of sound, the acoustic disturbance causes a non-linear, three-dimensional shockwave cone which trails the trajectory of the bullet. This leaves traces of an acoustic shockwave and shockwave reflections visible within the captured waveform. Exceeding the speed of sound means shockwaves appear before the sound of the muzzle blast as they are the first wave to reach the microphone (Maher and Shaw, 2008).

Reflections

Although the direct blast is brief, the capture will include longer responses due to multipath reflections from the ground and obstacles within the environment, both of which can be categorised as the early reflections. There will also be late reflections caused by higher-order reflections (reflections of reflections). As such, this area contains information about the environment rather than the weapon and can, therefore, be useful in differentiating between locations from which shots were fired rather than any information pertaining to the identity of the weapon (Maher, 2011).

Recordings of gunshots may also include sounds relating to the trigger and other firearm mechanisms.

Analysis methods

Pre-processing

Before an examination can be performed, signals require preparation in the form of pre-processing to optimise any analysis scheduled to take place. In the case of gunshots, it is known that the peak spectral energy of a muzzle blast is between 500 Hz and 1000 Hz, while the energy of a shockwave is above 2000 Hz. Filtering and noise reduction can, therefore, be applied to mitigate the effects of broadband noise and increase the focus on the desired signal (Begault et al., 2019).

Critical listening

Critical listening, with audio enhancements and in authentication examinations, is usually the first stage of any acoustic analysis and is used to determine an overall approach and document areas which require further analysis using objective methods. It should not be used in isolation to render conclusions due to its subjective nature, and although experts can be trained to discriminate between sounds based on certain characteristics, evidence from so-called 'experts' who claim they are able to hear things that others cannot should not be accepted, as there is no such thing as a 'golden ear' (Begault et al., 2019).

Waveform analysis

This type of analysis is used to provide information about relative timings and signal magnitudes at the receiver microphone. In the case of gunshots, it can be used to discriminate between the locations of two shooters based on the peak magnitude of the shots. Due to the aforementioned convolution of acoustic impulse events caused by the environment and signal processing, it can be difficult to isolate events without being provided with any further corroborating information.

Envelope analysis

The envelope of a gunshot sound can be defined by splitting the window into short frames across the time domain, followed by calculating the root mean square of each frame. To define the bandpass envelope, a filter bank can be applied to the frames before the energy of specific bands is calculated. As envelopes can characterise the attack and decay of a signal, as well as the average peak energy, they can be used to identify multiple events by calculating the period between individual signal peaks. Envelopes can also be used to determine the consistency of various events with one another through overlapping and correcting to a common onset time.

FFT analysis

Fast Fourier Transform (FFT) analysis can be applied to short acoustic events such as a muzzle blast to provide its frequency composition and allow comparisons between multiple instances to be performed.

A spectrogram can provide further visual information with regards to the relationship of an event's temporal, frequency, and amplitude characteristics. For a gunshot, the spectrogram representation would be expected to be a short, high magnitude event which covers a broad range of frequencies. It will not provide any form of 'fingerprint' of a weapon due to the effects of the reflections within the environment, but it can be used to provide analysis of the differences between shots due to a change in position or azimuth in relation to the microphone.

When performing analysis using FFT, the parameters should be documented (such as the frame size, window function, and percentage of overlap between frames), as this can affect the frequency domain representation, and thus the results. When analysing multiple events, the same parameters must be used for each to ensure differences or similarities are representative of those captured within the signal and are not caused by the settings. As with all FFT analysis, considerations must be made with regards to the time versus frequency resolution, as, for example, a short frame length with no overlap will provide an excellent time resolution, but poor frequency resolution.

Cross-correlation

Cross-correlation is the most commonly used method for calculating the similarity of sounds in a quantitative manner. It does so by measuring the parity between a time-series signal and a time-shifted signal as a function of lag and is also the covariance (or measure of the relationship) of the two signals normalised by the product of their standard divisions. The correlation coefficient 'r' explains the percentage of variance between the signals where $r^2 = 100$, and results range between -1 and +1, where +1 is a perfect linear relationship and -1 a perfect linear relation with a 180° phase shift. A correlation of 0.5, therefore, explains 25 per cent of the variance between the two waveforms (0.5^2). Its advantage lies in the fact that it quantifies the differences caused by the source, propagation channels, and recording systems. Caution must be practised when comparing the same firearm at different azimuth angles, as the use of cross-correlation analysis has been shown to produce significantly different values, and the artificial shortening or smoothing of signals increases the cross-correlation (Lacey et al., 2014), as does a reduction in the bandwidth of the signal (Lacey, 2019). The latter is surmised to be caused by the cumulative effect of reduced high frequency variations, the effects of which were found not to compromise the ability to distinguish between gunshots from different firearms.

Summary

The application of acoustics to forensics is a particularly challenging area, due to the extensive number of variables caused by the combination of an uncontrolled capture environment and the recording device colouring the captured audio. Add to that the potentially unpredictable effects of high magnitude sound, such as that emitted from gunshots, on the encoding algorithms of devices and it should be clear that this is a field where caution is required, and presenting the limitations of any findings is of paramount importance. That being said, the results of acoustic examinations have provided extremely telling and significant information in the most difficult of cases, including the assassinations of US Presidents.

References

Beck, S.D., 2017. Physics and psychoacoustics related to forensic ear witness accounts. Paper presented at the International Conference on Audio Forensics, Audio Engineering Society, Washington, DC.

Beck, S.D., 2019a. Who fired when: Associating multiple audio events from uncalibrated receivers. Paper presented at the International Conference on Audio Forensics, Audio Engineering Society, Porto, Portugal.

Beck, S.D., 2019b. A short-time cross correlation with application to forensic gunshot analysis. Paper presented at the International Conference on Audio Forensics, Audio Engineering Society, Porto, Portugal.

Beck, S.D., Nakasone, H., and Marr, K.W., 2011. An introduction to forensic gunshot acoustics. The Journal of the Acoustical Society of America 130, 2519.

Begault, D.R., Beck, S.D., and Maher, R.C., 2019. Overview of forensic audio gunshot analysis techniques. Paper presented at the International Conference on Audio Forensics, Audio Engineering Society, Porto, Portugal.

Freytag, J.C., Brustad, B.M., and Brustad, P.E., 2005. A survey of audio forensic gunshot investigation. Paper presented at the 26th International Conference: Audio Forensics in the Digital Age, Audio Engineering Society, Denver, Colorado, USA.

Koenig, B.E., Hoffman, S.M., Nakasone, H., and Beck, S.D., 1998. Signal convolution of recorded free-field gunshot sounds. Journal of the Audio Engineering Society 46, 634–653.

Lacey, D.S., 2019. The effects of bandwidth reduction on cross-correlation computations in the analyses of recorded gunshot sounds. University of Colorado, Denver, CO, USA.

Lacey, D.S., Koenig, B.E., and Reimond, C.E., 2014. The effect of sample length on cross-correlation comparisons of recorded gunshot sounds. Paper presented at the 54th International Conference: Audio Forensics, Audio Engineering Society, London.

Maher, R.C., 2007. Acoustical characterization of gunshots. Paper presented at Signal Processing Application for Public Security and Forensics Conference, IEEE, Washington, DC.

Maher, R.C., 2010. Directional aspects of forensic gunshot recordings. Paper presented at the 39th International Conference: Audio Forensics Practices and Challenges, Audio Engineering Society, Hillerød, Denmark.

Maher, R.C., 2011. Acoustical modeling of gunshots including directional information and reflections. Available at: www.montana.edu/rmaher/publications/maher_aes_1011_prezo.pdf

Maher, R.C., 2016. Gunshot recordings from a criminal incident: who shot first? The Journal of the Acoustical Society of America 139, 20–24.

Maher, R.C., 2019. Shot-to-shot variation in gunshot acoustics experiments. Paper presented at the International Conference on Audio Forensics, Audio Engineering Society, Porto, Portugal.

Maher, R.C. and Hoerr, E.R., 2018. Audio forensic gunshot analysis and multilateration, Paper presented at the 145th AES Convention, Audio Engineering Society, New York.

Maher, R.C. and Routh, T.K., 2015. Advancing forensic analysis of gunshot acoustics. Paper presented at the 139th AES Convention, Audio Engineering Society, New York.

Maher, R.C. and Shaw, S.R., 2008. Deciphering gunshot recordings. Paper presented at the 33rd International Conference: Audio Forensics – Theory and Practice, Audio Engineering Society, London.

Van der Kloot, W., 2005. Lawrence Bragg's role in the development of sound-ranging in World War I. Notes and Records of the Royal Society 59, 273–284.

Chapter 9

Speech forensics

Introduction

Although considered outside the scope of this book due to the prerequisite of expertise in other fields, for the sake of completeness, this chapter provides an overview of speech forensics. The reader interested in learning more is encouraged to seek out specific content relating to this area, as its reach extends far beyond digital audio forensics and into statistics, digital signal processing, and phonetics.

Although the terms 'speaker identification' and 'speaker recognition' appear to be used interchangeably by media outlets and the public at large, they are, in fact, two distinctly different areas. The term 'speaker comparison' encompasses both, as each relates to the comparison of voices.

Nolan (1993) defined the term 'speaker identification' as: 'an utterance from an unknown speaker has to be attributed or not, to one of a population of known speakers for whom reference samples are available.'

For a speaker comparison exercise to be performed, reference samples from a relevant population of speakers with voices that contain similar characteristics to the offender, in the eyes (or ears) of an expert should be used. All of the subfields which come under the speaker comparison umbrella are defined by their names, in which the first word pertains to the data type it is applied and the second to the task performed. The following is a summary of the main sub-disciplines.

Speech recognition is a term used to refer to systems which decode the actual words of which speech is composed. It can be applied to audio recordings or calculated in real-time, for instance, in speech-to-text applications. It generally has no use in forensics so it will not be discussed further.

Speaker verification is used to compare the voice of a known speaker who is willing to cooperate against that of an unknown speaker and is, therefore, a binary decision (either it is the correct speaker, or it is not). This type of system has applications for security purposes, for example, identity checks when calling a bank. Again, this has little use in audio forensics.

Speaker identification is used when the identity of a speaker is unknown, and so rather than verify the voice against that of a known speaker as in the previous subfield, anybody within a relevant population could be the speaker. The difficulty of this task is increased by factors such as:

- distortion caused by clipped samples;
- non-contemporary voice samples which may be several years old;

- speakers who are unwilling to cooperate in providing samples of their voice;
- open-ended data sets (you can never be sure that the criminal is among the suspects).

All of the terms describing the task in the aforementioned areas (recognition, verification, and identification) imply both a probability that the suspect and offender are the same people, and that a categorical result (where a finding is reported as one of several pre-defined categories) can be given. This is something which is not logically possible. Accordingly, it is argued by some experts in the field that the term 'forensic voice comparison' should, therefore, be used instead of 'forensic speaker comparison' as:

1   It defines the assessment of similarity between suspect and offender samples with regard to given parameters and commonality of those parameters within the wider relevant population. Relevancy is crucial when performing any comparison exercise, as to compare an English-speaking male against a Spanish-speaking female would be an obvious mismatch that would result in unreliable conclusions caused by statistical bias.
2   The objects of comparison are voices, and not speakers (Rose and Morrison, 2009).

History

The forensic analysis of voices dates back to the 1960s, made possible by the spectrogram technology developed during the Second World War. The term 'voiceprint' was first coined by Grey and Kopp (1944) through their research into the use of spectrograms while at Bell Labs. The intention of the research was to use spectrograph representations to extract identifying features of a voice, similar to identifying features within fingerprints, hence the term 'voiceprint.' When the war ended, the requirement and drive for this type of research declined until the New York Police Department made contact with Bell Labs in the early 1960s with a request for help in relation to cases involving phone calls in which the identification of suspects was sought (Smith, 2006). An employee of Bell Labs led the efforts to further the research and develop a methodology for a spectrographic voice by publishing his research (Kersta, 1962). It was claimed the method had an error rate below 1 per cent, a statement which was challenged to some degree by other scientists who all reported much higher error rates of between 18–50 per cent when attempting to reproduce the experiments.

Based on further research, the use of this approach was summarised by (Hazen, 1973), who concluded, 'The data suggest that the value of spectrograms for speaker identification purposes is limited to use as an investigating aid, and then only if speech samples of similar context and adequate duration are compared.'

Although the FBI set up a laboratory to focus on speaker identification (and reported low error rates over an extended period), they do not present their findings in court and use it for investigative purposes only (Koenig, 1986). Other methods have since been developed, although spectrogram analysis remains one of the available approaches.

Variables

As there are many factors which can cause variations relating to voices and their capture, it is essential these are understood and controlled, and error mitigation steps implemented where possible. In the first instance, there are a number of variations relating to both the voice (known as intrinsic variations) and to the capture channel (extrinsic) variations. The intrinsic variations are all related to the behaviour of a speaker during speech production, for example, the words spoken, and their mental or physical state at the time of capture. When you consider the majority of audio recordings used in a forensic context are captured during an event (an obvious example being a criminal act of some description), the speaker may be physically exerting themselves in some manner, such as running or fighting. They may also be intoxicated. They are also likely to be under some form of stress, which could cause further deviation from the characteristics of their voice under normal conditions. Extrinsic variations relate to effects after the speech has been produced, such as variations in the method of capture and the recording conditions (Drygajlo et al., 2016). For example, questioned recordings are typically from a phone call and captured in less than ideal conditions. In light of this, they often have low signal-to-noise ratios, contain overlapping speech, reverberation, and distortions. As telephones have a low bandwidth and use codecs for compression, aspects of the speech signal are also artificially attenuated (Künzel, 2001).

Invariably, the suspect recording used for comparison will be taken in controlled conditions, for instance, while the suspect is in police custody, and will, therefore, be of a much higher quality than that of the unknown recording, resulting in a quality mismatch (Morrison, 2009).

Many of the extrinsic issues which cause degradation to the quality of the voice can coincide, and if processing is applied to improve the overall quality of the recording, it should be used with extreme caution, thus ensuring no alterations are made to the captured voice which could lead to unreliable results.

Further to this, intrinsic variations can result in differences between the speech produced by the same speaker (known as intra-variation). For analysis to be conducted, these potential differences must be understood as not to be attributed to being from a different speaker and causing a false negative. The differences between two different speakers, caused by biological factors, such as vocal tract length, and social factors, such as age, sex, regional background, and ethnicity, are known as inter-variations (Morrison and Thompson, 2016).

Analysis methods

Forensic voice comparisons consist of two stages: the analysis, and the reporting of the results of the analysis. There is currently no consensus between those within the voice comparison community with regards to the method of analysis and reporting of results and, thus, there are a number of approaches for both stages. According to (Morrison et al., 2017), analysis methods are traditionally grouped into four areas, which he defines as follows.

Auditory

This method, which at face value appears to be the easiest, is also the one which requires the most experience to be performed to a high level of reliability. During the analysis stage, the expert listens to the recordings of each voice and comments on the properties which are similar and those which are considered unusual, including features such as dialect, laryngeal activity, and speech impediments. Basic acoustic measurements are also used before all of the findings are considered, and a conclusion is reached. As the majority of features are qualitative (grouped based on the examiner's perception rather than an objective measurement), they cannot be quantified, and thus reproducibility is difficult.

Acoustic phonetic approach

This approach requires taking quantitative measurements of the acoustic properties of the voice (such as formants and the fundamental frequency) before analysis of the numerical data is performed. Formants are defined as the resonant frequency of a vowel that reflects the articulatory configuration, shape, and size of the vocal tract. In order to perform the measurements, comparable units are first defined, such as phonemes, allophones, or stretches of speech. *Phonemes* are distinctive classes of actual speech sounds formed by *phones*, and the term *allophones* is the collective name given to all phones (Pulgram, 2015). These sounds provide a basis for comparison of speech. For instance, acoustic and phonetic comparisons can be made between two vowels which are part of the same phoneme, such as the vowels in the words 'car' and 'far' (Rose, 2002).

Measurements can also be made periodically, for example, by calculating the fundamental frequency and formant frequencies of regions which contain speech over regular intervals. The measurements for this technique are made using software under human supervision.

Spectrographic

This method involves the visual inspection of the spectrogram, which can also involve listening (which would then be an auditory-spectrographic approach). In the first instance, a section of the speech is selected, which could be a word or a phrase, before comparisons are made between the same word or phrase of the known speaker and a set of foil speakers. In order for this to be achieved, the suspect (and a set of foil speakers) must read the same phrase verbatim, using the same vocal effort, speech rate, and other factors to ensure the recordings are as similar as possible. To obtain intra-speaker variability, different phrases and repetitions of the same phrases are collected. Comparisons of various visual aspects within the spectrogram are then performed, including the general formant shaping and position, the energy distribution, pitch striations, and word length, followed by a comparison of audible features, such as pitch, stress, and word rate.

The subjective nature of visual analysis has been replaced in recent years by computers which take objective measurements of the spectrogram. Conclusions are stated on a subjective posterior probability scale with gradations such as probable identification, possible identification, inconclusive. Posterior odds refer to the probability of the prosecution hypothesis versus the probability of the defence hypothesis given the evidence, whereas *a priori* odds refer to the probability of the prosecution hypothesis versus the probability of the defence hypothesis, prior to taking the evidence into account (Morrison, 2014).

Automatic

The automatic approach draws on technology and methods used for automatic speaker recognition used in such industries as security and banking. The approaches share the methodology for the analysis stage, but a reporting stage (which is not necessary for speaker recognition purposes) is added for forensic voice comparison tasks, so findings can be presented in the form of a likelihood ratio. The analysis stage involves the taking of quantitative measurements, but instead of phonetic units as used in the automatic phonetic approach, features such as Mel Frequency Cepstrum Coefficients (MFCC) are used. These are based on the nature of human perception of frequencies by transposing the frequencies to the 'Mel Scale,' which is linear below 1000 Hz and logarithmic above this (Tiwari, 2010). Conversion to this scale is performed by measuring the spectrum in a 20–30 ms frame before the frame moves in 10–20 ms steps across areas in which speech is present. This process produces a series of sequential MFCC measurements. Other techniques use higher-level features based on phonetic units, such as phenomes, word or phrase length (Morrison, 2013).

The advantage of this approach is that it is fully automatic and so eliminates bias, is objective, can be validated and provides results in a likelihood ratio format. The examiner's role is limited to only the input of recordings and the reading of the result at the output.

Reporting methods

There are a number of methods using for reporting the results of forensic voice comparison analysis, and according to results of a survey performed by Gold and French (2011a), there is currently no general international standard or consensus. The following are the most common reporting methods used.

Binary decision

A binary decision allows only two possible outcomes, either they are the same voice, or they are not the same voice.

Classical probability scale

This reports the probability of the unknown and known voice being from the same speaker, and is typically a verbal scale which includes a number of steps such as 'highly likely to be the same voice/highly unlikely to be the same voice.'

Likelihood ratios

It is widely agreed within the forensic voice analysis that Bayesian likelihood ratios are the most appropriate expression for reporting the results of a voice comparison examination. Likelihood ratios provide a gradient estimate of the strength of the evidence based on the probability (p) of the evidence (E) given (|) the prosecution hypothesis (H_p) to the probability of the evidence given the defence hypothesis (H_d) (Gold and Hughes, 2014).

The odds form of the likelihood ratio is defined as follows:

$$LR = \frac{p\left(E \mid H_p\right)}{p(E \mid H_d)}$$

As the scale is centred around 1, any likelihood ratios above this support the prosecution hypothesis, while ratios below it support the defence hypothesis. As the result is a ratio, it represents how much more likely the evidence would be, given the prosecution hypothesis as opposed to the defence hypothesis. For example, a likelihood ratio of 5 should be interpreted as the evidence being five times more likely to be obtained assuming the prosecution hypothesis (the unknown and known voices pertain to the same speaker) than assuming the defence hypothesis (the unknown and known voices pertain to a different speaker).

UK position statement

In the UK, a new framework for presenting voice comparison evidence was proposed by a group of speech scientists after concern over the impressionistic likelihood ratios (e.g. 'highly likely to be the same person') that were in use at the time. The framework sought to bring the method for reporting results in line with other forensic science disciplines and was circulated to all forensic speech scientists within the UK for their feedback and thoughts. The result was all but one individual co-signing the proposed framework, and thus, a new method of reporting was born (French and Harrison, 2007). Rather than make identifications, an opinion is given as to whether the characteristics of the voice in the questioned recording match those of the voice of the suspect. The ultimate decision as to whether the voice does, in fact, belong to the same speaker is left to the trier of fact.

The UK reporting scale is composed of two parts. First, an assessment as to whether there is any support for the two voices belonging to the sample speaker is performed. If it is determined that there is support, a further opinion as to how distinctive or common the voices are is then reported. If is it determined there is no support that they pertain to the same speaker, then a conclusion of 'no support' is rendered (Gold and French, 2011b).

Summary

Although the analysis of speech primarily involves the analysis of the human voice, it also requires expertise in a range of other disciplines due to the scope for misinterpretation caused by the large number of variables pertaining to its capture. One of the most prominent is that of digital audio, as it is this medium that stores the pertinent voice data, and only by having an understanding of this field can artefacts associated with digital audio be detached from the true characteristics of a human voice. Forensic Voice Comparison examinations consist of two stages, the analysis and the reporting of results, and although research has shown that there are a number of techniques which are consistently used by those performing voice comparison tasks, there is no overall consensus in the use of these techniques or the reporting of results pertaining to the findings from the analysis stage.

References

Drygajlo, A., Jessen, M., Gfroerer, S., Wagner, I., Vermeulen, J., and Niemi, T., 2016. Methodological guidelines for best practice in forensic semiautomatic and automatic speaker recognition. Verlag für Polizeiwissenschaft, Frankfurt am Main.

French, P. and Harrison, P., 2007. Foreword to 'Position statement concerning use of impressionistic likelihood terms in forensic speaker comparison cases.' International Journal of Speech Language and the Law 14(1).

Gold, E. and French, P., 2011a. International practices in forensic speaker comparison. International Journal of Speech Language and the Law 18, 293–307.

Gold, E. and French, P., 2011b. An international investigation of forensic speaker comparison practices. Paper presented at the 17th International Congress of Phonetic Sciences, Hong Kong.

Gold, E. and Hughes, V., 2014. Issues and opportunities: The application of the numerical likelihood ratio framework to forensic speaker comparison. Science & Justice 54, 292–299.

Grey, G. and Kopp, G., 1944. Voiceprint identification (Report). Bell Telephone Laboratories, New Jersey.

Hazen, B., 1973. Effects of differing phonetic contexts on spectrographic speaker identification. The Journal of the Acoustical Society of America 54, 650–660.

Kersta, L.G., 1962. Voiceprint-identification infallibility. The Journal of the Acoustical Society of America 34, 1978.

Koenig, B.E., 1986. Spectrographic voice identification: A forensic survey. The Journal of the Acoustical Society of America 79, 2088–2090.

Künzel, H.J., 2001. Beware of the 'telephone effect': The influence of telephone transmission on the measurement of formant frequencies. Forensic Linguistics, 8(1), 80–99.

Morrison, G.S., 2009. Forensic voice comparison and the paradigm shift. Science & Justice 49, 298–308.

Morrison, G.S., 2013. Tutorial on logistic-regression calibration and fusion: Converting a score to a likelihood ratio. Australian Journal of Forensic Sciences 45, 173–197.

Morrison, G.S., 2014. Critique by Dr Geoffrey Stewart Morrison of a forensic voice comparison report submitted by Mr Edward J Primeau in relation to a section of audio recording which is alleged to be a recording of the voice of Dr Marlo Raynolds. Available at: forensic-evaluation.net/raynolds/ Morrison - critique of Primeau report - Raynolds case

Morrison, G.S., Enzinger, E., and Zhang, C., 2017. Forensic voice comparison. In I., Freckelton and H. Selby (eds), Expert evidence. Thompson Reuters, Sydney, Australia.

Morrison, G.S. and Thompson, W.C., 2016. Assessing the admissibility of a new generation of forensic voice comparison testimony. Columbia Science and Technology, Law Review 18, 326–433.

Nolan, F., 1993. Auditory and acoustic analysis in speaker recognition. In J. Gibbons (ed.), Language and the law. Longman, London.

Pulgram, E., 2015. Phoneme and grapheme: A parallel. WORD 7, 15–20.

Rose, P., 2002. Forensic speaker identification. Routledge, London.

Rose, P. and Morrison, G.S., 2009. A response to the UK position statement on forensic speaker comparison. International Journal of Speech Language and the Law 16, 139–163.

Smith, J.M., 2006. The accuracy and consistency of spectrographic analysis for voice identification. Paper presented at the 121st AES Convention, Audio Engineering Society, New York.

Tiwari, V., 2010. MFCC and its applications in speaker recognition. International Journal of Emerging Technologies. Available at: www.researchtrend.net/ijet/ijet/4_Vibha.pdf

Vanderslice, R., 1969. The 'voiceprint' myth. Center for Research on Language and Language Behavior. University of Michigan, Ann Arbor, MI.

Chapter 10

Audio enhancement

Introduction

Audio enhancement is the term used to describe the procedure of improving the quality of an audio recording. It is crucial to the justice system, due to both the generally poor quality of audio relied upon in legal cases and the potential these recordings have for providing compelling evidence.

As discussed in previous chapters, forensic audio recordings are a world away from those produced in recording studios. It would be a rare occurrence to have a recording which is to be used for legal purposes that would not be improved to some degree by enhancement processes due to the less-than-ideal capture conditions. The only reason that comes to mind in which a recording would not require enhancement would be in the event of a crime in a recording studio live room or vocal booth while the recording was taking place.

Once introduced as evidence, the objective and easily accessible nature of the contents of a recording, combined with the fact that it captures an audio snapshot of a moment in time, mean it can be extremely compelling. As evidence can influence the outcome of a case, and people's lives are often irreversibly affected by said outcomes, every grain of information must be extracted from each exhibit. In the case of audio evidence, this means ensuring recordings relied upon by the justice system are of the highest quality possible.

Quality versus intelligibility

For the limitations of enhancement to be understood, the different methods for measuring the success of such must first be understood. Although in a general sense, enhancement is considered to be applying processes to improve the audio recording, how can we be sure it has improved? As the final destination is the human auditory system, the only real method to measure this factor is through subjective listening. Although some objective information can be obtained from the file specifications (such as sample rate and format), and statistics in relation to the content (such as peak value, average RMS (root mean square)), good specifications do not ensure a good quality recording, just as poor specifications do not necessarily equate to a poor one. An uncompressed recording of 96 kHz, 24 bits and a peak value of -3 dBFS (decibel full scale) captured by a device in the pocket of an individual in a public environment containing both loud music and background chatter would most likely capture nothing of evidential value.

On the other hand, a perceptually encoded recording of 8 kHz and a peak amplitude value of -20 dBFS captured via telephone may be highly intelligible. It is, therefore, essential that quality judgements are made based on the content rather than the specifications. Although there are some algorithms which can be applied to audio which correlate highly with Mean Opinion Scores (in which panels of listeners score audio on a number of factors), the results are meaningless without a reference point. For example, one of the algorithms, Perceptual Evaluation of Speech Quality, or PESQ (Rix et al., 2001, 2000; Telecommunication Standardization Sector of ITU, 2001), can be used to determine the quality of a transmission method, such as a telephone capture system, by calculating the PESQ of the transmitted signal in relation to the original (or reference). When working in forensics, there is no reference recording available upon receiving the audio exhibit, so quality assessments are based solely on an examiner's opinion. Although, in theory, the algorithm could be applied by an examiner to determine the success of an enhancement by using the original version as a reference point, in practice, the inferior quality of recordings reduces the reliability of the results. The use of this kind of algorithm in method validation and verification is promising due to its objective nature, repeatability, and low impact on financial and human resources, but its success is as yet untested.

Although the words quality and intelligibility are often used interchangeably when referring to recordings, they are not necessarily equivalent, so judgements have to be made in relation to each in isolation.

Quality – the subjective opinion of an individual in their preference for one recording over another, or how pleasing the speech is to them (Naruka and Sahu, 2015). The definition of pleasing is left to the listener.

Intelligibility – This is defined by International Organization for Standardization, (2001) as 'a measure of effectiveness in understanding language.' It can be measured as the percentage of words correctly identified relative to the total number of words within a recording.

Although related, they are fundamentally different. For example, VOIP (Voice Over Internet Protocol) is of high quality thanks to a generous bandwidth but has low intelligibility due to the potential for words to be dropped during transmission as the data is transmitted in packets. POTS (Plain Old Telephone System), which is slowly being replaced by VOIP systems, is of higher intelligibility due to its reliability, but the quality is lower due to its restricted bandwidth (Datta et al., 2018).

By way of an example, if there are two recordings, one captured by telephone with a broadband noise throughout, and a second, identical recording, with short, high frequency beeps and transient clicks throughout, the first would be deemed to be of a higher quality than the other while the intelligibility would be the same.

Research has shown that intelligibility is strongly associated with formants, and although the first formant alone was found to be only a minor contributor, there was a strong correlation between the second formant and speech intelligibility. Different bands also contribute differently to the intelligibility of words, with the 1.5–3.5 kHz bandwidth providing the most information (Alves et al., 2010).

Table 10.1 Factors effecting sound quality

Stage	Factor
Capture	Microphone response and performance
	Location and alignment of the microphone
	Microphone directivity
	Presence, level and spectrum of background noise
	Room reverberation
	Sound reflections
	Distortion
Processing	Equalisation and signal processing
	Dynamic signal processing
	De-noising
Reproduction	The frequency response of loudspeakers
	Location of loudspeakers
	Calibration of systems
	Reverberation time/reflectively of listening/playback room
	Presence of reflecting surfaces near loudspeaker and listener
	The distance of listener to loudspeakers and axial position
	Background noise level
	The sound level of audio playback
	Distortion of the audio production system
Talker	Articulation and clarity of speech
	Modulation
	Speech rate
	Accent
	Familiarity/fluency of language
	Voice level
Listener	Hearing acuity
	Attention/alertness
	Familiarity or fluency of language

There is a wide range of factors which impact on intelligibility, each of which can be broken down into the position in the chain between capture and playback at which they are introduced (Mapp, 2016) (Table 10.1).

As forensic audio examiners, we have no control over the stages relating to the capture process or the talker, but can control the processing (by following a framework and understanding the impact of each method), the reproduction equipment (for example, using headphones to eliminate the majority of issues relating to the reproduction stage), and the listener stage (by looking after our hearing, taking breaks, and performing repeat listening).

A recording deemed to be of high quality may not necessarily be of higher intelligibility than a recording of lower quality, as there are so many factors which can influence such. A recording captured in ideal conditions, but containing a competing speaker, may be of lower intelligibility than a low-quality recording captured in poor conditions but of a single speaker. In making judgements on quality, an examiner with experience will base their opinion on factors such as the number of artefacts and the signal to noise ratio. In terms of intelligibility judgements, this would be a subjective opinion on the number of words which can be accurately transcribed relative to the entire recording. Although generally it would be expected that good quality recordings will be of higher intelligibility than low-quality ones, there are exceptions.

Although the reason for most parties instructing an audio enhancement is to improve the intelligibility of the voices within, in theory, this is extremely difficult, for two reasons:

1   Various research studies have found that enhancement processes are designed to improve audio quality and not intelligibility. If you consider the removal of clicks from a recording, the method replaces the click through interpolation, utilising data from each side of the region in which the click occurs. This process is not designed to improve intelligibility as it is interpolating the entire spectrum, so any speech data in the interpolated region will be lost. It could be argued that normalisation can improve intelligibility, as speech may not be intelligible on account of its magnitude being too low. This can be true, but more often than not, and especially when captured at lower resolutions, low magnitude speech does not contain enough data to be intelligible, regardless of any post-processing applied. Normalisation also increases the level of the noise floor, and adds another stage of quantisation error, so in performing this process, the speech to noise ratio will be the same, or may even increase.

2   The human auditory system is highly effective in differentiating between speech and surrounding noise. We are living audio enhancement software packages, as shown with our ability to focus on a single conversation within a room full of speakers (known as the cocktail party effect). Although we lose some of the cues which are available in the real three-dimensional world, such as Interaural Time Difference (ITD) and Interaural Level Difference (ILD), our brains are still extremely adept at focusing on speech content within a recording. Thus, if enough data pertaining to speech has been captured, the actual quality level of the recording has less impact on our brain's ability to interpret it than one may think.

In the 1970s, Lim performed research into the intelligibility of enhanced audio and found there was no improvement (Lim, 1978). Some 30 years on, research by Hu and Loizou found none of the enhancement algorithms available at the time of testing improved the intelligibility of speech relative to unprocessed speech, apart from a small improvement (<10 per cent) in recordings captured in vehicles (Hu and Loizou, 2007). This is surmised to be due to the stationary nature of the type of noise emitted from vehicles which lends itself to being accurately estimated by noise reduction algorithms. It can, therefore, be removed more effectively while not affecting the data relating to speech. As non-stationary and unpredictable noise such as speaker babble cannot be tracked, it cannot be accurately estimated, and, therefore, cannot be removed with the same degree of precision.

The danger of applying noise reduction processes are two-fold. First, the noise is removed, but the remaining speech may become distorted as a by-product of the processing. Second, the residual noise left in the signal becomes more speech-like, leading to interference with speech information, potentially changing the meaning of words.

Measurement methods

As there are two different domains in which audio can be represented, approaches to each have been proposed as the result of various research studies. Overall it has been found that a time-domain signal to noise ratio (SNR) measurement is poorly correlated with both quality and speech intelligibility. In contrast, frequency-domain SNR correlates well, as it uses critical band frequency spacing to mimic human frequency selectively, as well as weighting functions for each band.

The two most common speech intelligibility measurements are AI (Articulation Index; French and Steinberg, 1947; Kryter, 1962) and STI, or Speech Transmission Index (Steeneken and Houtgast, 1980). AI is a refined version of the Speech Intelligibility Index (American National Standards Institute, 1997), and is based on the fact that intelligibility of speech is in direct proportion with the amount of frequency information that the listener can perceive. This is calculated by dividing the spectrum into 20 spectral bands and estimating the weighted average of the SNRs in each band (based on band functions). It can predict the effects of both linear filtering and additive noise but does require the input of both the speech and masking signal to be taken at the eardrum of a listener, an option which is not generally available when working with recorded audio. It has been validated for stationary masking noise (based on long-term spectra or 125 ms intervals), but not moving noise (Ma et al., 2009).

The vast majority of enhancement processes multiply the original signal by a suppression function based on specific error criteria such as MSE (mean squared error). MSE is widely used in a number of applications to calculate the actual value of a signal by providing the average of the squares of the differences between actual and estimated values of each sample (Gray and Davisson, 2004).

Although multiplication is used as a basis for enhancement processes, it can also cause two types of distortion:

- *spectral attenuation*, in which the enhanced spectrum components are smaller than their corresponding input components;
- *spectral amplification*, in which the enhanced spectrum components are larger than their corresponding input components.

The above are not distinguished from one another in the previously mentioned measurement tools to calculate estimations. PESQ (Telecommunication Standardization Sector of ITU, 2001) addresses this by applying different weights for these distortions and is the recommended methodology for the speech quality assessment of 3.2 kHz handset telephony and narrowband speech codecs. The result is calculated as follows:

1   The original and degraded signals are equalised to a standard level and filtered with the frequency response of the handset.
2   The signals are time-aligned to correct for time delays and processes.
3   The loudness spectrum is obtained through an auditory transform.
4   The difference in loudness is computed before being averaged over time and frequency to produce a prediction of a subjective quality rating between 1.0 and 4.5 (in which a higher value represents better quality audio).

This method has shown a high correlation (0.92) with Mean Opinion Scores in a range of conditions using VOIP (Rix et al., 2001), and 0.9 using noise-corrupted speech processed via noise-suppressing algorithms (Hu and Loizou, 2008).

For these reasons, it is the current standard, but its limitation lies in the requirement for both the clean and processed signal. It is, therefore, of no use determining the quality of an exhibit. Its potential application is for the testing of enhancement processes and as a subsequent quality assurance stage in casework by comparing the original version to that generated.

Mean Opinion Scores are those which are given subjectively by a panel and scored on a scale of 1–5 (International Organization for Standardization, 1996). There are several methods to do this, all of which involve the reproduction of speech through a communication system to listeners within a controlled environment. The listeners must then answer questions relating to the audio. The ANSI/ASA (2009) standard has listeners opine on monosyllabic words produced by loudspeakers over the communication system being tested. Another is called 'rhyme testing' in which rhyming single-syllable words are produced in random order (Fairbanks, 1958). This was later improved in what was named the Modified Rhyme Test (House et al., 1965). All of these methods are used to eliminate context, thus preventing the panel of listeners from using clues from surrounding content.

The problem with Mean Opinion Scores is that they are expensive due to the use of subjects to both produce the words and to participate in the listening exercise. The procedure also requires the use of facilities suitable for the testing process. In addition, certain elements have been shown to be unreliable (Huckvale and Frasi, 2010), as noise-reduced signals were found to achieve higher scores despite their lower intelligibility. This serves as another clear indication that quality and intelligibility should be evaluated separately, and evidence that poor quality audio can be highly intelligible and vice versa.

Two further methods for the measuring of audio quality/intelligibility are as follows:

1   Segmental SNR. This is a frame-based approach performed by calculating the average SNR estimates for each frame across the length of a signal in the time domain. One of the problems with this measurement is that the estimations become very negative during periods of silence as the signal energy is small, and, thus, so is the difference between the desired signal and noise (Hansen and Pellom, 1998). As the correlation between SNR and subjective quality is low, this method is of limited use.
2   SNR_{ESI}. This is the frequency domain version of segmental SNR, and has a correlation score with MOS of 0.81. Its success is due to the use of bands to model the frequency selectivity of the human auditory system through the application of a perceptually weighted function to replicate these bands.

As processes are designed to improve the quality of recordings, the aim of an enhancement (as it relates to speech) should, therefore, be to improve the quality of a recording while maintaining the intelligibility by not altering (through removal or addition) any data associated with the speech. It may be the case that intelligibility appears to improve as a by-product of the quality improvement, but to focus solely on improving the intelligibility of the speech would make no sense when the processes are all designed for improving the *quality* of audio.

Preparatory analysis

The enhancement of audio recordings is no different than the authentication aspect in that the most original recording available is always desired. As it relates to enhancement, this will ensure it is of the highest quality possible as it will not have undergone any other processing or format conversion since capture. Requests should, therefore, always be made to the instructing party for bit-stream copies of the most original versions (Koenig et al., 2007). This should be done as soon as possible as, depending on the circumstances, the organisations involved, and the general nature of the case, it can take time. Before any enhancement processing is performed, an initial review (much like a mini authentication) of the exhibit should be undertaken through the analyses of the metadata and spectrogram to determine whether it is consistent with an original recording. Obviously, a full authentication, although desirable, is not possible for every recording provided to the laboratory due to resource limitations, so the caveat should be made within the report that a full authentication examination was not performed as it is beyond the scope of an enhancement. It is the expert's duty to enhance the audio to the highest quality possible, and this often involves advising those instructing (who generally have little understanding of audio) of the exhibits required in the first place and reasons for them.

There is a myriad of factors which influence the quality of a recording and the potential of an enhancement, and for a successful enhancement, a plan must be formulated by first understanding the issues and then documenting how to reduce or remove the impact on quality that these issues may have. A good starting point is to document the format and specifications of the recording. Although, as highlighted in Chapter 9, there is no direct correlation between a recording's specifications and the quality or intelligibility, the cause of any issues can be understood. For example, if the recordings are perceptually encoded and there are audible birdies, as we know the origin of such artefacts and the frequency region they occupy, we can easily determine how best to remove or reduce them without impacting any of the speech data.

The second stage is to review the audio in both the temporal and spectral domain to document issues which can be determined visually, such as differences in speaker level (in the time domain) and magnitude variations of specific frequency regions (in the spectral domain). This is best paired with the critical listening stage, using a similar methodology used during authentication, but noting quality issues rather than artefacts.

The following describes the issues which should be documented during a critical listening examination before an enhancement is performed.

Background noise

Background noise is easy to discern as we deal with it every minute of our lives. Any sounds which are not correlated with the desired signal can be considered to be background noise, and the type, regions occupied, and general magnitude in comparison to the desired signal should be documented.

Clicks

Clicks and pops in audio are generally defined as undesired audible transients caused by the system, and not the acoustic environment (Sørensen, 2005). The listener perceives them in many ways, but in the realm of digital audio, they are often heard as tiny tick sounds caused by poorly concealed digital errors and timing problems within the ADC (analogue to digital converter) of the capture device (Godsill and Rayner, 1998).

Transient noise

Transient noises are those which are relatively short in nature, such as birds chirping or an alarm clock tone. The times at which they occur should be noted, as well as the type of sound.

Continuous tones

Continuous tones are those present throughout the entire recording or for an extended period, such as the hum emitted from lights, or electromagnetic interference from mobile phones. Again, the times at which the tone begins and ends should be documented, thus allowing for further investigation with regards to their frequency content during the second stage of analysis.

Competing speakers

The presence of competing speakers is common during recordings captured in public venues, such as bars and outdoor spaces where people gather. Unfortunately, they are one of the most challenging issues to solve during enhancement as all speech occupies the same frequency bandwidth and their transient nature makes them impossible for machine learning algorithms used in noise removal to predict. Any regions in which there are competing speakers should be documented for further investigation.

Broadband noise

Broadband noise is most prominent on recordings in which the desired sound source within the environment is of low magnitude, but the gain setting of the microphone is high, such as recordings made when the microphone is in a different room to the sound source. It is audible as a white noise-type sound which occupies the entire frequency range and is caused by the quantisation error of the ADC becoming audible due to the sparse acoustic content in that region (due to either a lack of sound or sound of a low magnitude).

Reverberation

Forensic recordings are often recorded in less-than-ideal spaces with little to no thought of microphone placement. Add a reverberant environment into the mix, and the result can be a loss of transients and the smudging of words due to early and late reflections, reducing the intelligibility of speech. The acoustic reverberation of speech can be described mathematically as:

$$x(n) = s(n) * h(n)$$

where $s(n)$ represents the clean speech signal and $h(n)$ the impulse response of the environment (Cole et al., 1995). The characteristics of reverberation are modelled as linear systems, and the impulse response of this system is dependent on the room's acoustic properties as well as the position, direction, and frequency response of the microphone (Mahieux and Marro, 1996).

Reverberation becomes increasingly noticeable as the time between the direct and indirect signals (known as pre-delay) extends. Generally, the indirect signal will be of a low magnitude and thus its impact on the desired signal negligible, but when captured in an environment constructed using highly reflective, solid materials (such as custody cells or an interview room), the reverberation can become both pronounced and problematic. The opposite is true for voices in small, highly absorbent spaces, such as cars, which are padded with a large amount of soft material. The presence and degree of reverb should be documented, and if this changes at any point during the recording, information pertaining to such should also be noted.

Speaker levels

The magnitude of speech captured within a recording can vary based on several factors, including:

- gain setting of the microphone;
- the distance of the speaker from the microphone;
- the direction of the speaker in relation to the microphone;
- on-axis position of the microphone;
- any occlusion of the microphone diaphragm (for example, within a pocket);
- any occlusion of the speaker's mouth;
- environment (sound generated outdoors will dissipate spherically, whereas sound captured indoors will be both reflected and be absorbed);
- level of effort on behalf of the speaker (whispering, talking, shouting).

The levels should, therefore, be documented, in terms of general overall magnitude and the dynamic range. It may be that there is a consistent amplitude difference between two or more voices, for example, during telephone calls captured acoustically at one end of the call. One reason may be the voice of the subject at the end of the line with the capture device being of higher magnitude than that transmitted over the telephone.

Distortion

Distortions can come in many forms and can have a plethora of causes, such as the encoding algorithm used, the transmission channel, and the occurrence of clipping. It is relatively easy to differentiate between general noise and distortion, as the distortion will be correlated with the desired signal, and so will follow the pattern of such. Distortion is also more likely to occur when there is a sound of high magnitude which exceeds the available quantisation level values of the system, resulting in clipping and the creation of a square wave. If the causes are apparent, these too should be documented.

Table 10.2 Critical listening checklist

Category	Feature	Present (Y/N)	Offset	Notes
Distortion	Clicks/crackle			
Distortion	Clipping			
Distortion	Mobile phone burst			
Distortion	Reverberates			
Distortion	Signal loss			
Distortion	System noise			
Distortion	Aliasing noise			
Distortion	Transmission Interference			
Distortion	Non-linear distortion			
Periodic noise	Tones			
Periodic noise	Sirens			
Non-periodic noise	Hum			
Non-periodic noise	Convolutional changes			
Non-periodic noise	Environmental noise			
Non-periodic noise	Speaker level differences			
Non-periodic noise	Hiss			
Non-periodic noise	Broadband noise			
Non-periodic noise	Coughs, steps, etc.			
Source separation	Background music			

Overall signal magnitude

The overall volume of the recording should also be considered. In order for an opinion on such to be consistent throughout all examinations undertaken, the gain setting of the digital to analogue converter should remain in the same position at all times. If it is adjusted at any point, it should be reverted to its default position once the examination has concluded. The level can be further verified through visual analysis of the waveform and the calculation of amplitude statistics such as RMS level and peak amplitude.

Documentation

Table 10.2 is an example of a form which can be used to document the results of the critical listening procedure.

Enhancement suitability

Once the quality issues have been documented, a judgement should be made as to the suitability of the exhibit for enhancement by reviewing the requirements of the client and assessing the potential for meeting these requirements. Not all audio is suitable for enhancement, and even where it is, there can be certain limitations. These should be relayed to the client to both manage

Table 10.3 Comparison of workflows of studio enhancement, manipulation, and forensic enhancement

Studio enhancement	Manipulation	Forensic enhancement
1. Record musicians 2. Mix multitrack session to improve the overall enjoyment of sound 3. Export recording	1. Record audio 2. Remove, clone or splice regions 3. Export recording.	1. Create a bit-stream image of the original exhibit. 2. Document and compare hash checksums. 3. Preliminary analysis of the recording. Document findings. 4. Compose methodology for enhancement based on the analysis. Document. 5. Enhancement of the recording, ensuring no pertinent data is lost. Document enhancement processes and settings used. 6. Export enhancement. Document hash checksum of the generated exhibit. 7. Write a report, documenting a variety of information, including limitations and methodology.

expectations and allow them to make an informed decision as to whether to instruct the work to proceed.

Manipulation versus enhancement

The reasons for maintaining the integrity of exhibits have been covered in earlier chapters, so will not be repeated here, but it is vital to understand how this concept relates to enhancements, as the audio is technically being manipulated. Unlike an authentication examination, in which data is extracted from the recording, enhancement requires the audio to be processed, and in doing so, the end product differs from the original version (or working copy of such).

To understand the difference between a forensic enhancement, a recording studio enhancement, and a manipulation, the general workflow of each scenario is presented in Table 10.3.

Presenting the stages in this manner makes clear the fact that documentation is an integral aspect of forensic audio enhancement. As with all forensic processes, this ensures the process is repeatable and reproducible. If there are questions raised over the work performed at a later date, the documentation of the process undertaken can be provided for another examiner to review, and, if required, repeated. This is not the case for studio enhancements, and definitely not for nefarious manipulations. Even in cases where the client may think that a report is unnecessary due to the nature of the case, contemporaneous working case notes should be always be taken as one can never be sure where the enhanced evidence will end up.

As quality issues often affect specific local regions within a recording, there is no magic button (or process) that can be applied which will optimally enhance each section. In general, enhancements consist of two stages: identification of the issues, followed by the application of processes to reduce or remove the issues. It is the reduction or removal of the impact of these issues that results in the overall enhancement. Most quality issues have at least one corresponding solution, and some may be approached using one of a number of different tools depending on the recording and the preference of the examiner. Degradation can be thought of as any undesirable modification to an audio signal, and can be classified in two ways, according to Mathai and Deepa (2015):

1   *local degradation*: Only affects a particular group of samples, for example, clicks, crackles and clipping.
2   *global degradation*: Affects all samples, for example, broadband and background noise.

Quality issues and solutions

The aim of this section is to provide information pertaining to the various issues encountered within audio recordings, and the possible solutions. Each is presented to a level of detail to ensure an understanding of what causes the issue and how the enhancement process can assist from a technical standpoint. It is only through understanding how the methods work that the most suitable settings can be selected, and an optimal enhancement achieved. The enhancement of an audio recording is analogous to the excavation of a fossil. It may be possible to determine what the fossil is, but it takes great skill and many tools to remove the redundant material surrounding it, using great care to ensure the desired content is not damaged. It is the shedding of this excess material that produces the end result.

Clipping

Clipping is the result of an input signal exceeding the dynamic range capabilities of the capture device's ADC to store information relating to the magnitude of the signal. Its occurrence results in modifications of both the amplitude and frequency distribution of the original sound. For example, if the maximum quantisation level is 65536 in an unsigned (non-bipolar) 16-bit system, any input that exceeds this upper limit will be represented as 65536 as this is the maximum level available. Although clipping is generally associated with the upper limits of a signal, it can also occur at the lower end, when an input is so quiet that it does not meet the lower limit. In this case, the sample is quantised as zero, regardless of how low the magnitude of the signal is (Gaspar, 1992).

Clipping can be classified as either 'soft' or 'hard.' Soft clipping occurs when input values above the clipping threshold are driven back down, but still represent the general distribution of the input signal, albeit at a reduced magnitude. This is found in the analogue domain, where all clipping is soft, and thus causes a harmonic distortion effect. Hard clipping occurs when values above the clipping threshold are set to the maximum value available within the system, resulting in a flat cut-off that can cause audible high-frequency harmonics (Figure 10.1). Clipping in the digital domain is always hard and causes non-harmonic distortion (Mathai and Deepa, 2015).

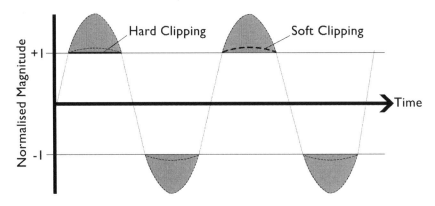

Figure 10.1 Hard and soft clipping in the time domain

Cause

Common reasons for clipping are setting the input level of the capture device at a level unsuitably high for the sound being captured, the occurrence of an unexpected loud signal within the capture environment, or using a bit depth that is too low to accommodate the type of recording being captured. It is difficult to prepare for events that are unexpectedly loud, such as gunshots, but with preparation, clipping can be prevented in all but the most unpredictable situations. The root cause can be an input which was clipped (either at the microphone or ADC level), post-processing (such as an increase to the magnitude) and coarse spectral quantisation which takes envelopes into a higher value than the original input when using very low bit rates (Chalil et al., 2008).

The detection of clipping can also be used for authentication, for example, a signal which contains an area consistent with clipping, but is below 0 dBFS is likely to have been re-encoded as clipped samples would be expected to exist only at the maximum available quantisation level.

Clipping has become more of an issue with recordings captured using mobile phones, mostly due to the use of an unsuitably high gain setting while within proximity of a sound source.

Effect

The effects of clipping are two-fold. First, information over a certain threshold is not uncaptured, and so the input signal is not accurately represented. Imagine a sound with an arbitrary magnitude of two, but the capabilities of the system result in the sound being clipped to a value of one. Fifty per cent of the information pertaining to that sound is not represented by the audio recording, altering the general timbre of the sound. This loss of information can render tasks such as voice comparison unreliable (Ramirez, 2016).

Further to this, the encoding of a number of sequential samples generates a square wave, in which the shelf is present at the maximum value the signal can represent. This square wave causes an audible distortion which becomes more noticeable to the human auditory system the longer the number of consecutively clipped samples extends, resulting in masking and a generally unpleasant sound. The combination of both this and the former point are most easily identified in loud music within close proximity of the microphone, as not only does a large amount of data pertaining to the music go uncaptured (causing a coarse representation of the sound), but distortion is also clearly audible due to the introduction of a square wave. As there are transitions at either end, it is not a pure square wave and does not fulfil sampling theorem (which would require an infinite bandwidth), so results in aliasing. The distortion itself is caused by the rounded and oscillating shape of the reconstructed square wave upon application of the low pass anti-aliasing filter upon playback of the signal (Kraght, 2000).

Distortion can be categorised as follows:

Linear: Changes to the relative amplitude and phase of the frequency components present in a complex signal are perceived as changes to the timbre or tone quality.

Non-Linear: Components are introduced which are not present in the original signal, often perceived as a harsh or rough character to the sound and the introduction of crackles and clicks.

(Tan and Moore, 2003)

Solution

Clipping should always be removed as the square waves created can cause distortion, which impairs both fidelity and audio quality. Processing that takes place with the clipping present may also produce further distortions as the input of a sequence of repeated samples will be replicated at the output if the algorithm is not based on feedback.

Clipping is removed in two stages, namely, detection, followed by replacement of the clipped samples. First, the areas which are clipped must be identified (Laguna and Lerch, 2016). This is accomplished through either histogram visualisation or time-domain analysis and can be performed visually or automatically. Visual methods include waveform analysis (time-domain) and histogram analysis (the occurrence of clipping will result in large tails to the distribution due to the truncating of the amplitude of the signal). The histogram method works when the signal is uniformly clipped, but its effectiveness is reduced for signals where there are only a few areas of clipped samples (Aleinik and Matveev, 2014). The time-domain methods work by detecting flat slopes and discontinuities at the beginning and end of the clipped region, visible as sharp increases and drop-offs to the sample values. Some methods compare the waveform against the peak level, based on the theory that sections closest to the peak level are more likely to be damaged than those closer to the centre of the waveform. Other algorithms define clipping as an area of three consecutive samples at the maximum quantisation level, although this is very conservative as clipping under 2 milliseconds is likely inaudible. Riemer et al. (1990) applied a differential that detects clipping when the derivative is above a certain threshold, and the signal amplitude is close to the maximum or minimum available value.

Next, the frequencies present before and after the clipped area are analysed, and the section bridged accordingly by the reconstruction of the waveform. The available methods fall into three categories: time domain, frequency domain, or sparse reconstruction. The time domain techniques work through interpolation or extrapolation, where samples are created based on those either side of the clipped area (Janssen et al., 1986). Frequency domain methods work on more extended periods of clipping where the time domain method is insufficient due to the lack of data from which to interpolate new samples (Lagrange and Marchand, 2005). Sparse reconstruction methods vary and are computationally expensive. One approach ignores clipped samples and interprets the signal as being sampled at non-linear intervals before the best sparse representation is obtained (Nielsen and Lund, 2003). The audio is then divided into bands and analysed through the peak-to-RMS average ratio for each band. These dynamic detectors set the ratio of multiband upward expanders (which cause increased expansion when fewer dynamics are present), while the threshold is kept at a fixed offset from the peak level of the original audio in each band. The decompression caused by the upward expanders prevents peak-limiting and compression in the densest regions of the recording.

De-clipping should never be used for an enhancement which is to be subsequently used for voice comparison tasks as the interpolated regions are not representative of the real voice and may be misinterpreted during the analysis, leading to inaccurate and unreliable results. Finally, in relation to enhancement processes, reducing the overall level of the entire signal through the use of gain or normalisation will likely prevent further, unintentional clipping due to the enhancement processes, but will not remove the square wave as no repair process has been applied to clipped samples. The distortion will, therefore, remain, albeit at a lower magnitude.

Clicks

Clicks can be defined as finite defects which occur at random amplitudes, for random durations (typically between 1 and 200 samples at a 44.1 kHz sampling rate), and at random positions in the waveform (Figure 10.2). Due to their short duration, they do not add any information to a recording, so are generally considered to be undesired transients. Clicks can have a similar effect to clipping and can be additive (added to the desired signal) or can replace the signal entirely (potentially resulting in masking) for a short period. Although clicks and pops are technically noise, as they are not correlated with the signal, they have their own special category due to their extremely short length and the method used to remove them. The difference between a click and a quick burst of general background noise is essentially subjective and determined by the examiner. With some experience, the maximum duration of a click which can be effectively removed using de-clicking algorithms can be determined, and if the removal causes distortions or is ineffective, it would not be considered a click for all intents and purposes. For example, the click of a computer mouse may be easily removed due to its short duration, but the clicking sound made by a pen hitting a table may contain reverberation, resulting in a longer signal that cannot be removed using the same method.

Cause

Clicks can be caused acoustically by a short transient within the environment (such as a pen click or microphone pop from speech), but more often than not the source is non-acoustic. For instance, switching noise, communication system errors, signal dropouts, physical surface degradation (to physical storage media such as CDs and DVDs), poorly concealed digital errors, and timing problems within the ADC of the recording device.

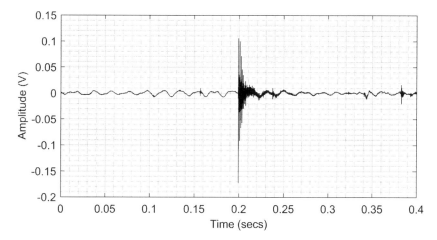

Figure 10.2 Single click event in the time domain
Source: Created using Mathworks Matlab.

Effect

Clicks are classed as a type of impulse noise (or impulse disturbance) and can be perceived in various ways, from tiny tick noises through to scratches and crackles. Not only can they be distracting, but in extreme cases may also cause the masking of phenomes, reducing speech intelligibility. Although interpretation is an unavoidable aspect of forensics, limiting the use of it should be pursued where possible. The unmasking of phonemes has a critical role in increasing the intelligibility of a word and can lead to a reduction in the level of subjectivity required in the interpretation of speech.

Clicks can also reduce the effectiveness of other enhancement processes. Imagine a recording that contains non-simultaneous low magnitude speech and a single high magnitude click. If peak normalisation is applied, the gain increase will be limited by the click if the level of the click is below the level set for normalisation. If the peak level of the click is higher than the level set for normalisation, it could actually result in a reduction of the level of the low magnitude speech. If gain is applied with a focus on the speech content with no consideration for the click, it will most likely result in clipping and the associated distortion.

Solution

There are various techniques used to remove clicks, all of which follow the familiar pattern employed by clipping removal, namely, detection followed by reconstruction. The more simplistic techniques make the assumption that no useful information is within the samples containing the click and interpolate the entire region, while advanced techniques extract information from the degraded samples using noise modelling. An effective method can be considered to be one in which interpolation across regions of up to 100 samples at a 44.1 kHz sampling rate go undetected by the listener.

The ultimate aim, as with all enhancement techniques, is for only artefacts audible to the listener to be removed while not affecting the desired signal in any way. Any unnecessary processing risks distorting the perceived signal quality and removing critical data, so a balance must be sought between the impact of current issues pertaining to the quality and the artefacts which may be introduced during processing.

There are a number of methods, ranging from simple to complex, so only a few will be covered. These can be summarised as follows (Betts et al., 1992):

SAMPLE AND HOLD

This method samples the signal before the click event and holds (repeats the sampled signal value) beyond the area containing the click. This technique is based on the theory that a plateau will be closer to the desired signal than a click. A high pass filter is used for the detection of clicks, which makes the assumption that most signals contain little information at high frequencies, while impulses contain spectral content at all frequencies. It employs a user-defined amplitude threshold to locate any transients which appear above the threshold. This is effective to some extent, but leaves audible bumps and pops in the waveform, and does not work if the signal has a large amount of high-frequency content or has been low pass filtered.

LINEAR INTERPOLATION

Linear interpolation corrects the clicks by measuring a sample before the click (as with sample and hold) and a sample afterwards, then essentially forming a gradient of sequential samples between the two samples. The audible result is less intrusive than the sample and hold technique, but causes low-frequency artefacts and bandwidth reduction over the area. Dither of the interpolated area is usually applied by the algorithm to reduce quantisation error (and subsequent quantisation noise) of the corrected region.

MEDIAN FILTERING

Median filtering is one of the classic approaches, and replaces the corrupted samples with median values of the samples either side while retaining the detail in the waveform of the signal, but is of no use for clicks over a few samples in duration (Godsill and Rayner, 1998).

SPLICING

This technique splices uncorrupted data into the degraded section and ensures there is no discontinuity between the good and replaced samples, taking advantage of the periodic nature of speech (Goodman et al., 1986).

SPECTRAL REPAIR

One of the more novel methods is to reconstruct the area in which the click resides through analysis of the samples which occur pre- and post-click. Due to the high order of interpolation, this method is undetectable to the human ear but runs the risk of changing phonemes, potentially resulting in misinterpretation of speech content. The majority of software which employs this method is proprietary, and as such, the sample level processing that occurs is unknown. It can be logically deduced that the unwanted area is removed and then resynthesised using the surrounding areas, similar to the de-clipping process, but is essentially a black box process of which little is known with regards to the method for reconstruction of the region.

Regardless of the method used, the area containing the click should be defined with the utmost degree of precision possible to limit the amount of data changed, a concept which should always be practised when performing forensic audio enhancements. It also reduces the potential of audible artefacts caused by the de-clicking process. If artefacts are introduced when the process is applied with a high degree of precision, the click is likely too long for effective removal, or the regions surrounding the click are of a quality which renders them unsuitable for sampling for the interpolation of the area in which the click resides. In these cases, other techniques to reduce the perception of the click must be sought.

Ambient noise

Ambient noise can be defined as acoustic sounds which occur within the capture environment that are recorded by the capture device but do not form part of the desired signal.

Cause

As forensic recordings are most commonly captured in uncontrolled conditions, it is likely that there will be a number of extraneous sounds within the recording which do not correlate with the desired signal, are not caused by the system, and are not of a duration small enough to be considered a click. These may include sounds from passing vehicles, voices, screams, ringtones, and emergency services vehicle sirens.

Effect

The effects of ambient noise are wide-ranging, due to the multitude of sound sources which can cause it, and the various characteristics of which they are composed. These extend from those that are merely distracting to the listener, through to the simultaneous and complete masking of the desired signal. The intensity of captured speech can also depend on the intensity of the ambient noise in which the voice is immersed (French and Steinberg, 1947). In light of the sheer number of variables, the noise may only occupy a small bandwidth within the frequency spectrum, for example, the classic telephone ring tone, or cover the spectrum in its entirety, such as that produced by vehicle engines, exhausts and tyre noise. It may exist within a small region within the time domain (think of a text message alert tone) or may be present for the entirety of the recording, for example, voices within a restaurant.

Solution

Although there are a number of solutions, the variations in the character of this type of noise mean each must be considered on merit, and the most suitable solution applied. The primary objective should always be to ensure whichever process is used does not remove any data pertaining to the desired signal. The second objective is to remove the maximum amount of unwanted noise data. It is in finding the perfect balance between these two factors that the optimal enhancement will be achieved.

One of the most common forms of noise reduction is equalisation. The term itself is used for processes which target the spectral content of a recording and change the magnitude of frequencies in relation to one another. The name itself stems from the use of fixed response equalisers to boost high frequencies that were initially lost over long telephone transmissions to ensure the magnitude of signals output by the transmitters at each end of the line would be equal. The first application of equalisers for improving sound post-capture was when an RCA employee named John Volkman used variable equalisers in cinemas to improve the quality of their playback systems in the 1930s (Bohn, 1988). Nowadays, they are generally associated with digital audio and are employed to manipulate the frequency spectrum in order to achieve the desired sound. In audio forensics, they are used to attenuate areas that are of detriment to the desired signal and boost areas which render an improvement in the desired signal. This may be to improve the clarity of speech through boosts to high-frequency areas, or improve the overall quality of the audio through attenuation of lower frequencies caused by wind hitting the diaphragm of a microphone. A digital filter is an algorithm that convolves input samples with an impulse response and outputs the resultant samples. In doing so, the input spectra are multiplied with an ideal filter characteristic in the frequency domain (for example, a low pass filter), which is equivalent to convolving the time-domain signal with the impulse response of the time-domain. Equalisation processes are also non-destructive, so can easily be removed from the signal chain, allowing the signal to return to its pre-equalisation state. A non-destructive process should always be chosen over a destructive process when working with forensic audio to ensure as little of the original signal is removed as possible.

There are generally two categories used to describe an equaliser. The first relates to whether it is static (also referred to as stationary) or variable (also referred to as dynamic). Static filters work by attenuating or boosting frequencies specified by the filter settings for the entire duration of the recording, much the same way that the original equalisers for telephones would be used. They are, therefore, ideal when the spectrum of the region to which the filter is being applied remains relatively constant throughout. Where static equalisation processes function in a linear manner based on predefined settings and are independent of the audio input, a variable equaliser reacts, based on both the pre-defined settings and the audio input, much like a compressor. Rather than performing compression on selective frequency bands like a multiband compressor, dynamic equalisation uses traditional EQ in place of the gain reduction, giving much more control over the signal (iZotope, 2014). Where the terms 'variable' or 'dynamic' are not explicitly stated, it can be assumed that the equaliser is static in nature.

The second category used to describe an equaliser is the region of the spectrum it affects. For instance, the term 'low pass' indicates only the lower frequencies are allowed to pass from the input to the output, and 'band stop' indicates a band of frequencies are stopped from passing from the input to the output. The terms filter and equaliser are used interchangeably as an equaliser 'filters' audio.

When putting the two descriptive categories together, both the type of filter and region it affects are defined, for example, 'dynamic bandstop' filter or 'static high pass' filter. As the type

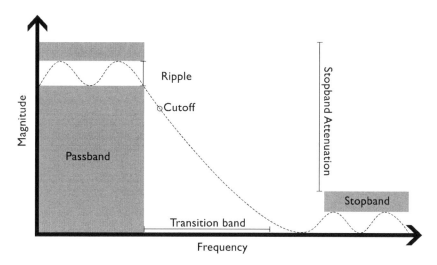

Figure 10.3 Low pass filter

of filter required for each quality issue encountered may differ, the following section provides a survey of the filters available.

LOW AND HIGH PASS

Low pass filters attenuate the level of the signal above a user-defined cut-off frequency (defined as the point at which -3 dB of attenuation occurs), thus letting lower frequencies below the cut-off pass through to the output unaffected (Figure 10.3). High pass filters are the opposite and attenuate signals below a defined frequency, and thus let the high frequencies pass (Figure 10.4). A user can define the slope (the gradient between the cut-off frequency and the point at which maximum attenuation is reached) in decibels per octave (dB/octave). Common values are multiples of six, so a 6 dB filter would transition gently, whereas a 30 dB filter would take only two octaves to reduce the signal by 60 dB, which is effectively muting (Izhaki, 2008). In configuring the settings, a balance should be sought to achieve the steepest gradient possible to accurately remove all unnecessary data while minimising artefacts that may be introduced in doing so. The most prominent artefact related to filters is that of 'ringing,' so named as it creates a ringing-type sound around the cut-off frequency.

SHELVING FILTERS

Shelving filters can be considered the less brutal relatives of low and high pass filters, as they use a reference frequency rather than cut off a frequency, which leaves one side of the spectrum undisturbed while applying attenuation or boosts to the other side at a consistent level (hence the use of the word 'shelving'). The magnitude of the boost or attenuation is definable by the user as

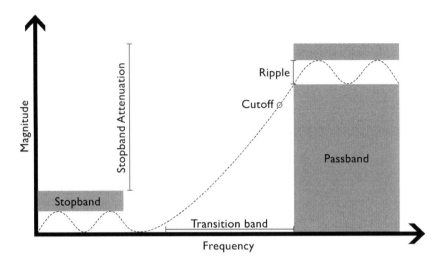

Figure 10.4 High pass filter

gain. These filters are useful when a subtle change to the spectrum of a recording is desired, for example, attenuating low frequencies caused by the microphone proximity effect, or applying a small amount of boost to upper band frequencies to improve the clarity of speech.

PEAKING FILTER

Peaking filters are much like a surgeon's scalpel in their precision, allowing a user to apply boosts or attenuation to small frequency ranges focused around a selected centre frequency (Figure 10.5). The quality factor (Q) defines the width of the curve, where a high Q indicates a narrow band, and a low Q indicates the opposite. The Q is the inverse of the bandwidth, in which the bandwidth is the range of frequencies that lie between the upper and lower -3 dB points on the curve. Q is calculated by dividing the centre frequency by the bandwidth (Huber and Runstein, 2005). As they can target specific frequency bands, they are ideal for the reduction of static sounds, such as hum and other stationary tones. They are also sometimes referred to as notch filters.

Settings available to the user are the same as the comb-filter (which is a series of notch filters) applied at equal periods within the frequency range.

DYNAMIC EQUALISATION

When noise is static (or constant) in its frequency composition, a standard static filter is capable of removing the noise to some degree to improve the quality of the desired signal. If, however, the noise is varying rapidly or has a complex spectrum, a filter which can adapt to the changing noise profile is required. Adaptive noise filters work by predicting future samples from previous samples, which can be especially effective with repetitive low-frequency noises such as wind, hum,

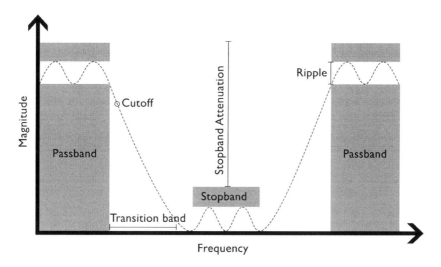

Figure 10.5 Band stop filter

and engine noise. The presence of speech that competes with the desired signal is more difficult to remove due to its unpredictable nature (Betts et al., 2005). The number of coefficients used for the processing is defined by the number of 'taps,' the number of which is usually user-definable. The more taps used, the higher the frequency resolution, meaning narrower filters with steeper roll-off rates. The downside to this is the extra processing power required. Adaptive filters are also of no use in removing unpredictable transient sounds such as clicks, as previous samples give no indication that a click is to occur in the future. Even with good material, care must be taken to ensure speech data is not processed, which can be indicated by a thinning in its pitch. By using a dynamic EQ, the signal remains unprocessed in regions containing no noise, thus reducing the amount of processing required while maintaining the spectrum of the desired signal.

Filters are ideal for the removal of excessive frequencies which have no relation to the desired signal, for example, a low pass filter removing everything above 6 kHz within a speech recording. Not only does this enhance the desired signal by removing noise, but it also serves to refine the processes which follow by allowing them to focus on the desired signal and not waste resources on unwanted noise.

Broadband noise

Broadband noise refers to a signal of random amplitude encompassing the entire range of frequencies within the audio spectrum.

Cause

The most common cause of broadband noise is quantisation noise becoming audible. Once a recording device is set to record, it begins taking samples of the capture environment. If the environment is tranquil, such as in an empty room, then the sample values to represent the environment will be of an extremely low level. As the quantisation noise of a recording has had dither applied to cause an increased variation between the values, the result is a sound with a character comparable to white noise. As the ratio of signal to noise is dependent on the bit depth, increasing the gain of the device will only increase the level of the quantisation noise. It is for this reason that broadband noise will be most prominent within low magnitude recordings with a high microphone gain. That being said, most recordings contain some form of broadband noise, which will be most evident during periods of silence within the acoustic environment being captured.

It can also be caused by acoustic events, such as wind, which appears primarily in the low-frequency range, but naturally extends across the entire spectrum (Brixen and Hensen, 2006).

Effects

At a low level, broadband noise can reduce the overall quality of a recording as it usually covers the entire spectrum, and often the entire length of a recording, assuming it was captured in the same environment with the same gain setting applied throughout.

At a higher magnitude, not only does it become more distracting, but can also mask lower level signal components. This can also affect further processing, such as de-reverberation. As reflected signals are always lower in magnitude than the direct signal, and may be partially masked by the broadband noise, the effectiveness of the process is reduced if this is not dealt with first.

Broadband noise is most intrusive at high frequencies for two reasons. First, this area generally contains less acoustic data, so the noise is not masked as it is in other regions. Second, even when high-frequency content is present, the perceptual masking effect is not as strong in this area.

Solution

The spectrum of a noisy signal is the sum of the desired signal and any undesired noise, and can be defined as:

$$X(n,k) = S(n,k) + N(n,k)$$

where $S(n,k)$ represents the spectral coefficients of the desired signal at bin n, frame k, and $N(n,k)$ represents the noise (Stolbov and Ignatov, 2012). The removal of noise from a signal is not an easy task as there is only a single known $X(n,k)$, and two unknowns (S and N). Approaches to noise reduction therefore consist of two stages, where the first is to determine which elements of the known signal are noise and which are the desired signal, and the second is to reduce the noise. The approaches for achieving noise reduction generally fall into two familiar categories, those which occur in the time domain, and those in the frequency domain.

TIME DOMAIN

Time-based techniques work by attenuating the signal whenever a region of samples is detected which contains only noise, triggered by a drop below a user-determined threshold. This is based on the assumption that if the signal is above the threshold, it is presumed to contain noise and the desired signal, and if below, it contains only noise. This gain reduction creates the perception to the listener that the noise level across the processed recording is lower than it is in the original. It is generally referred to as a noise gate or downward expander and includes user-definable settings such as:

- *threshold*: the signal level at which the attenuation is triggered;
- *attack*: the time it takes for the signal level to be reduced;
- *release*: the time it takes for the signal level to recover after reduction;
- *range*: the amount of attention applied to the signal once it drops below the threshold.

Some techniques feature a look-ahead option which will analyse the audio before the area being processed to ensure samples at the beginning of the region are not reduced due to slow attack times. There are a few limitations to this method. If the signal to noise ratio is high, even when the desired signal is present, it will do little to enhance the recording (Maher, 2005). It is also of no use if the desired signal is continuous in nature with no breaks. More often than not, noise will also exist in regions at the same magnitude as the desired signal, meaning this will go untreated and remain in the signal.

SPECTRAL DOMAIN

Spectral domain processing is a ubiquitous technique in restoring a signal that has been corrupted by broadband noise, and there are many methods to achieve the desired outcome. One of the more established techniques for the removal of broadband noise is that of spectral subtraction, first proposed by Steven F. Boll (Boll, 1979). This is the umbrella term under which the majority of spectral domain noise removal techniques reside. This method uses estimations of the short-term spectral magnitude of noise (at each frequency bin and for each frame) as a function of frequency, and subtracts it from the noisy signal, under the assumption that it is uncorrelated with the desired signal. Since it was first proposed, various improved iterations based on spectral subtraction have been published and implemented, including critical-band spectral subtraction (Singh and Sridharan, 1998), multi-band spectral subtraction (Kamath and Loizou, 2002) and dynamic spectral subtraction, all of which take advantage of redundancies in the human auditory system in relation to masking.

The pitfall of this technique is that any of the desired signal elements falling below the noise threshold will be removed with the noise, thus rendering this technique ineffective in extremely low signal to noise ratio situations. It can also cause various artefacts due to the statistical variance of short-time spectral estimates and random oscillations caused by calculated gains leading to bursts of energy in the processed signal, known as musical noise. A further problem is that if the signal does not contain any sections of isolated noise from which to capture a noise estimate (for example, if the recording includes overlapping conversations throughout the entire duration), the method cannot be used.

As broadband noise is prone to change throughout a recording, it may be that the estimated noise from a specific region is not representative of the noise throughout the recording. If the estimate does not match the region from which the noise is being removed, or the SNR is overestimated (as is often the case), it results in a signal mismatch. This can subsequently cause a birdie artefact (Lorber and Hoeldrich, 1997), where sinusoids vary in frequency from frame to frame and are audible as a whistling-type effect. Some of the more sophisticated algorithms incorporate functions such as constant updates to the noise estimate (for example, taking a sample of noise every 1000 samples to minimise any potential mismatches), not applying any subtraction when the desired signal is strong (as it is assumed the noise is already being masked), and suppressing any birdies which may have been introduced as a result of the processing.

Spectral subtraction can be a very effective technique but is highly dependent on the particular scenario and requirements. All currently known approaches in monaural recordings add distortion to the desired signal, and, as such, a compromise must be made between the amount of noise reduction and the amount of signal distortion introduced. Methods should be used to assess the results to ensure the optimal output for the exhibit recording has been achieved, including SNR measurements, critical listening, and FFT analysis.

As the impact of broadband noise is most audible at high frequencies, the simplest method is to apply a basic low pass filter. The caveat is that if there are other sounds in this range which are desired, these too would be removed. A dynamic low pass filter can be applied in these situations, which is inactive when the noise is masked, thus limiting the amount of noise reduction applied based on the temporal nature of the signal (Hicks and Reid, 1996).

Noise is a complicated quality issue to deal with as, if all noise is removed, some of the desired signal may also be lost. On the other hand, if some of the noise remains, the process may not as effective as it possibly could be. A balance should, therefore, always be sought between these two states.

A further problem is that it is impossible to remove broadband noise completely as:

1   The sum of two random signals always results in a loss of information about each of them.
2   Digital audio signals are quantised, which adds low-level quantisation noise to the signal, so any enhanced output will always contain some form of quantisation noise (Maher, 2005).

The second point can be negated by exporting to a high bit depth, thus increasing the SNR, something that will be covered later in this chapter.

Background music

Background music can be defined as any music which does not form part of the desired signal.

Causes

The cause of this issue is simple, the capture of music played within the environment of the recording device. This is an extremely common occurrence for forensic recordings as the majority of public meeting places (for example, bars, restaurants, and shopping centres) contain some form of background music, albeit at differing levels of intensity.

Effects

Background music is one of the most difficult elements encountered when performing an audio enhancement for several reasons. First, as it is highly transient and continually changing, it is not suitable for standard noise reduction techniques as the sound cannot be modelled. Newer, adaptive techniques are also unsuitable for similar reasons, as although certain genres of music are repetitive, the intro-variability is high. As such, previous samples cannot predict current samples, and by the time the system has adapted, the signal has changed.

Although time domain noise reduction techniques can be applied (dependent on the magnitude of the music), these can result in a distracting pumping-type sound as the attenuation can be perceptually obvious due to the continuous nature of the music and the dominant frequencies within which it resides. Music is recorded at a minimum sample rate of 44.1 kHz, so always occupies the entire audible spectrum of the human auditory system. When the desired signal is speech, a filter can be used to remove data which occupies frequencies outside of the desired area, but the remaining bandwidth (usually between 100 and 4000 Hz) will contain music across its entire range. To further compound the issue, if the music contains a vocal track, the desired speech and the vocals will convolve when captured, and as both occupy the same frequency bandwidth and both are transient, thus, separation is problematic.

A final issue is that music is often played continuously, especially in environments such as bars where tracks are played sequentially, with the ending of the previous track blending with the beginning of the current track. This leads to the sounds pertaining to such being audible throughout the entire capture.

It is often the case in law enforcement that background music or sounds emitted from televisions or radios will be encountered when dealing with covert recordings. Some individuals meet in public places which contain high levels of background noise in an attempt to sequentially mask what is being said, and may even purposely increase the amplitude of background music to mask speech, or at least make it difficult to transcribe. There are even devices available which replay pre-recorded conversations to mimic the ambience of a busy restaurant to mask speech within the capture environment.

Solution

Reference cancellation is the process of removing elements of a signal by taking advantage of phase differences between two signals. To serve as a simple example, if there is a musical recording featuring an artist singing, an exact copy of the music without the vocal can be subtracted, leaving just the vocal (Alexander and Forth, 2011).

As it applies to audio forensics, rather than isolate an artist's vocals, speech is generally the desired signal. In reality, it is not as simple as it first appears because forensic recordings will have spectral additions, such as reverberation from the room, the DAC responsible for playback of the music, and the ADC responsible for capture. These issues can be reduced through the use of a noise cancellation process, but it can often be painstaking in its implementation as a copy of the background music (the reference recording) must first be located and the music within each recording aligned with extreme precision. Both the sample rates and average spectra of the reference recording must also be corrected to match the evidence recording. In doing so, it is crucial that down-sampling rather than up-sampling is applied, as up-sampling would contain no new components above the Nyquist frequency, thus causing differences between the high-frequency spectra of the recordings. As reference recordings are often richer in low-frequency content because they have not been convolved with the room and the capture microphone, this too must be corrected through the use of manual filters or frequency equalisation algorithms (Ignatov et al., 2012). Recent years have seen an increase in software which can identify musical recordings (known as acoustic fingerprinting), which makes the task of finding the musical track much faster as it saves the time required for a manual search through a myriad of records to identify a track. This forensic technique is not just applicable to music but can be applied to any recording in which it is possible to obtain the reference track for, such as a television advertisement or a pre-recorded public announcement. It must be made clear that the content of the recordings must be accurately replicated in each recording for the method to be successful, so live music and speech are not applicable in this situation unless the music is being recorded through a direct line-in. Attempting to use a studio version to remove a live version would not be successful due to the nuanced differences between the two.

One method for removing the background music is through a landmark-based fingerprinting algorithm. It first locates frequency peaks (the landmarks) and determines the time differences between each landmark to create a 'fingerprint.' Original tracks within a database are then autonomously compared to this fingerprint to identify the song and start position of the music (Ellis, 2009). Signal cancellation is then performed using a normalised least mean square (LMS) algorithm.

Another method is the use of an adaptive filter with two channels, one containing the recording and another containing a reference. These filters compare both channels and determine which elements of the exhibit recording are from the reference and which are not. Similar to a standard adaptive filter, the process works through the prediction of future samples but does so from the reference track instead of the past samples of the exhibit recording (Alexander et al., 2012).

Hum

Hum is one of the more common noises encountered within forensic recordings, perceived as a humming or buzzing sound, and classed as an additive sinusoidal disturbance (Fechner and Kirchner, 2014). It has the character of a periodic signal consisting of a fundamental frequency and related harmonics, the magnitude of which depends on the cause of the hum due to the variability of mains voltages (Figure 10.6). The spectrum of hum is limited to lower frequency regions due to the small number of partial tones (or harmonics), whereas buzz contains a higher number of these tones, and subsequently disturbs a relatively large bandwidth (Brandt and Bitzer, 2014).

Cause

Hum is the result of disturbances caused by the mains power supply reaching the signal, during either the initial capture or through a subsequent capture (such as replaying the recording on a portable device while capturing it on a computer using a line-in cable). There are primarily two ways in which the mains supply can reach the signal. The first is through an inductive loop (when the capture device or cables are within an electromagnetic field), and the second through a

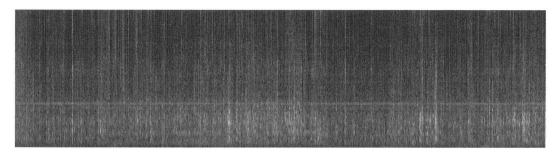

Figure 10.6 Spectrogram representation of a recording containing hum, visible as the lighter-toned horizontal line
Source: Captured from iZotope RX Advanced.

galvanic connection (when the capture device is directly connected to AC power). The root cause is usually faulty shielding of audio equipment or signal lines (Łuczyński, 2019).

One of the best-known sources of hum is ENF or electrical network frequency, which can be inadvertently captured if the device is mains-powered or in the proximity of a mains-powered device through electromagnetism. This phenomenon is caused by the alternating current (which is around 50 Hz in the UK and 60 Hz in the US) and the harmonics relating to this current. Although the harmonics exist at a lower amplitude, they can also extend into the bandwidth occupied by speech (Grigoras, 2009). Buzz most commonly occurs when recordings are captured in the vicinity of power lines or electrical lighting fixtures.

In electronic equipment, the DC voltage is obtained from the AC mains output using a step-down transformer and a diode bridge rectifier, which converts the energy to a full-wave rectification (so it has single polarity and flows in one direction, as opposed to AC, which alternates) before a low pass filter is used to hold the DC component of a signal using a large capacitor. The result of this process is three-fold:

1   The frequency of the output is twice as large as the input, so energy is sent twice as often.
2   The output is more abundant in harmonics.
3   The output contains a DC component for powering functional blocks of electronic equipment used for recording.

The culmination of these factors results in higher power from DC, causing more significant hum in the system (Dobre et al., 2015).

Effect

Low-frequency hum (below 100 Hz) can often go unnoticed by the listener due to the ear's limited response in this frequency region, but as the low-level frequencies contain a significant amount of power, the hum will de-optimise the algorithms used by processing if not removed. It may also obscure the true dynamics of the content of the sound, and thus affect any dynamic processing applied.

Higher frequency hum (including harmonics related to frequencies below 100 Hz) can extend into the spectra occupied by the human voice and beyond, so they must be reduced to prevent any simultaneous masking from taking place. Hum or buzzing which exists beyond the range of speech may not affect the intelligibility of speech, but will likely reduce the quality.

Solution

Similar to other enhancement techniques, the process must be performed in the sequence of detection followed by removal. Most methods estimate the fundamental frequency using a DFT algorithm to determine consistent peaks within the spectrum over an extended period, before filtering the signal to eliminate the unwanted component.

In terms of filter selection, a comb filter is generally the most effective solution. This consists of a series of notch filters set to attenuate the signal at linearly spaced intervals, removing the fundamental frequency and any associated harmonics relating to them. As the bandwidth occupied by each harmonic differs based on frequency, different settings of Q are more effective than a one-size-fits-all approach (Luknowsky and Boyczuk, 2004). As the harmonics of hum often fall below the average signal level once they exceed approximately 360 Hz (causing any audible perception of such to be masked), there is generally no need to extend the harmonics affected beyond this. If the frequencies are not linearly spaced, then adaptive filtering or manual notch filtering of the offending frequencies are the preferred options. Care should always be taken, especially in relation to higher-order harmonics which share bandwidth space with speech, to ensure the desired signal is minimally affected. This can be achieved through the use of accurate analysis when determining the bandwidth of the hum, combined with a high Q setting to ensure only a precise region is gently attenuated to prevent audible artefacts caused by bands of missing data. It is only necessary to reduce the magnitude of the hum to a point where it becomes inaudible. For hum restricted to low-frequency regions, applying a simple high pass filter may be all that is required.

The disadvantage of using a filter for hum removal is that areas of the desired signal may be lost in the process, even when using a band-stop filter with surgical precision due to the overlapping of desired information with the hum. As with all filters, the phase characteristics of the surrounding area may also change. Although a high degree of frequency selectivity may appear to be a good solution in limiting the degree of the desired signal removed, this can cause long impulse responses and resonance artefacts (Brandt and Bitzer, 2012).

An alternative method is to obtain a profile of the hum and then remove it using noise reduction algorithms, but again this has its disadvantages. In the first instance, a section of the recording which contains the isolated hum and no desired signal is required. Second, if there is a difference between the parameters of the extracted noise profile and those which need to be removed, the process will be ineffective. This would occur in a signal in which the frequency or amplitude of the hum changes over time.

Another novel method was proposed by (Łuczyński, 2019). The hum is first detected using DFT, followed by the synthesis and 180° phase shift of hum components relative to those contained within the recording. The synthesised signal is then added to the original version. This results in a phase cancellation (similar to that used in the removal of music) and was shown to be effective in reducing the level of hum with minimal inference or artefacts.

Reverberation

Reverberation is the term used for sounds reflected within an acoustic environment which have been captured during the recording process. When the desired signal is reflected, the associated reverberation is predictable as it is highly correlated with such.

The characteristics of reverberation are determined by the relationships between three components in the time domain. These can be defined as follows:

1   *Direct signal/path*: This is the signal emitted from the sound source, which travels directly to the microphone and thus is not reflected in any way. It reaches the microphone before any reflected elements because it has a shorter distance to travel, and is also higher in magnitude as it has not been absorbed by objects within the environment (which would attenuate the signal).
2   *Early reflections*: These are classed as reflections which reach the microphone approximately 50–80 ms after the direct signal and are caused by the room transfer function (RTF). They can be temporally and spatially discriminated as they are well-defined impulses of a relatively high magnitude. It is the early reflections which cause spectral coloration (modification of the frequencies of which a signal is composed).
3   *Late reflection/reverb tail*: Reflections which arrive at the microphone after the early reflections are classed as late reflections and are caused by the RTF coefficients. They are of a low magnitude due to the amount of absorption they have undergone, and are exponentially decaying reverberated energy. The reflections and diffusion from walls and objects within an acoustic environment make time and space distribution of impulse responses increasingly random. It is the late reflections which cause changes to the temporal waveform envelope.

Early and late reflections can be analysed to provide clues as to the size of the room and the distance between the microphone and source, but have a detrimental impact on both the quality and intelligibility of the desired signal as the direct signal becomes smudged by the reflected signals. The role of de-reverberation is simply to attenuate these reflections.

Cause

Recording systems have no way of differentiating between the source of sounds that enter a microphone, so both the direct signal and indirect reflections are convolved at capture. The characteristics of reverberation are modelled as a linear system, the components of which can be described mathematically as:

$$(n) = s(n) * h(n)$$

where $s(n)$ represents the clean signal and $h(n)$ the impulse response of the environment. The impulse response is dependent on the room size, acoustic properties (such as the absorption coefficient of walls and surfaces), as well as factors such as microphone position, direction, and polar pattern. It provides a complete description of a reverberant system, and it is the time of arrival

(TOA) and amplitude of reflections that determine the characteristics of this system. Acoustic spaces do not have a single room impulse response (RIR), so changes to the position of the microphone and sound source will result in a different RIR. As the relative positioning between the source and microphone position is virtually infinite, there is also an infinite number of impulse responses possible (Mahieux and Marro, 1996).

The biggest reverberation issue pertaining to audio forensics is we cannot know the value of $h(n)$ without the use of sensors or an array within the recording environment to estimate the RIR. As forensic recordings are more often than not captured in unknown or inaccessible (for a variety of reasons) environments, we generally do not have the luxury of knowing the $h(n)$ value.

Effect

Highly reverberant recordings are commonplace within forensics due to the uncontrolled capture environments. The effects of reverberation alone can render speech unintelligible before the numerous other issues that may be present within a recording such as distortion or noise are even considered. The reason for the adverse impact reverb has on speech data are twofold. The first is the build-up of the early reflections which arrive within 50–80 ms after the direct signal. Once they become louder than the direct sound, there is an internal smearing effect known as self-masking (Desmond et al., 2014; Sadjadi and Hansen, 2012). The second is caused by late reflections related to speech, causing a type of pre-masking known as overlap-masking (Bolt and MacDonald, 1949). This refers to the energy of the previous phoneme being smeared over time and overlapping the current phoneme. These tails have an exponential decay similar to that of the room impulse response.

As the indirect reflected sounds are delayed and attenuated versions of the direct signal, it is difficult to distinguish between the two in isolation. This results in a type of masking known as informational masking (as the masking is caused by the similar information contained within each signal). As it relates to the de-reverberation processing, the difficulties associated with separating parameters such as formants from room reflections is one of the biggest challenges in ensuring the direct signal goes unchanged (Kinoshita et al., 2008).

Another factor which affects the degree of intelligibility reduction is the speaker to microphone geometry. The further away the desired signal is from a microphone, the more smearing is caused by the acoustic reverberation due to the reduced magnitude ratio between the two. This is further compounded by the superimposition of multiple delayed and attenuated copies of the clean signal (Kodrasi et al., 2017). This reduction in intelligibility increases until the critical distance, at which point it remains constant. The critical distance is defined as the distance at which the direct path energy is equal to the energy of the combined reflections. For speech captured under the critical distance, intelligibility is dependent on the source to microphone distance, reverberation time, and the size of the room. Beyond the critical distance, it is dependent on the reverberation time only. Another point worth noting is that consonants play a more significant role in intelligibility than vowels (Habets, 2010).

Two of the most common perceptual effects of reverberation are the box effect and the distant talker effect. The box effect is defined as a reverberant speech signal perceived to be coming from different sources positioning at different locations within a room, thus arriving at different times and intensities. This adds spaciousness to the sound and makes the talker appear to be in a box.

Distant talker effect pertains to the perceived spaciousness (as above), making the speaker appear further away from the microphone than they are (Shujau et al., 2013).

Solution

Reverberation is more complicated than spectral subtraction or adaptive noise cancellation, especially when no information is available with regards to the acoustics of the capture environment. The term de-reverberation is used to refer to the identification and removal of reverberation from the desired signal using digital signal processing. De-reverberation is a form of convolved noise removal of which there are three categories, as follows (Gaubitch et al., 2010):

1 *Beamforming or spatial identification* – uses difference microphones which are weighted to form a beam of enhanced sensitivity in the direction of the desired source and to attenuate sounds from other directions (Veen and Buckley, 1988).
2 *Enhancement* – modification of the signal to improve the features of the clean speech.
3 *Blind deconvolution* – acoustic impulses are identified blindly, meaning there is no *a priori* information available with regards to the room impulse response or original anechoic signal. The impulses are then deconvolved to leave the original signal.

Research has resulted in a number of proposed methods, including cepstrum and signal property-based algorithms. Cepstrum-based techniques consist of an estimation of the RIR followed by equalisation. One of the earliest methods was proposed by (Oppenheim et al., 1968) in which simple echoes were observed as peaks in the cepstrum of the speech signal and removed using an algorithm to determine and subsequently attenuate frequency peaks with a comb filter. A second approach was to use a low pass weighted function applied to the cepstrum, under the assumption that energy is in the lower frequencies. Signal property algorithms work by locating intervals within small signal to noise ratio regions which are then attenuated.

Another common approach is to estimate the room response $h(n)$ before designing and applying an inverse filter $h'(n)$ to cancel out the room response. In order for this to be performed, the elements of the signal which are dry (a direct signal with no reverberation) and those which are wet (containing reverberation) must be determined. Most methods are a two-step procedure in which the room response is estimated, followed by the application of an inverse filter, either through least-squared error or cepstral separation techniques (Soulodre, 2010).

In a forensic context, there are a few factors caused by the recording process, which limit the success of de-reverberation. First, it is usually the case that the only information available is from a single microphone, which can mean cues such as the ITD and ILD are not available. Second, if no visual information such as video footage is available, clues that can be garnered from studying the synchronisation of speech with lips are also lost. Finally, and most crucially, forensic examiners always work blindly, that is to say, that the clean signal information is never available, and all that remains is an equation with two unknowns and a single known (the evidence recording signal). De-reverberation has many elements similar to that of blind source separation and spectral subtraction, whereas, in this instance, it is the acoustical impulse of the room and the desired signal for which separation is required.

There is a relationship between de-reverberation and blind source separation, in which both require the unmixing of a signal, with minimal information available. The term 'blind' refers to an algorithm that works without *a priori* knowledge of the signals or the mixing process it is trying to remove or isolate.

Another issue, highlighted in (Nees et al., 2016), is that many models assume a diffuse reverberation field, but rooms with non-uniform absorption caused by objects and materials of different absorption characteristics will have time-varying spatial properties.

Amplitude

Amplitude pertains to the magnitude of a recording, either globally (across the entire duration of a recording) or locally (within specific temporal regions).

Cause

There are many variables which can cause amplitude issues, of which the origin is not always obvious. Some of the primary reasons for global amplitude issues are the distance of the microphone from the source, the magnitude of the source itself, the gain setting of the microphone, and occlusion of the microphone (for example, having the recording device contained within a jacket pocket). Local issues can be caused by the differences in distance between two or more sources from the microphone, differences in the level of said sources, differences in the transmission channel of a source before reaching the microphone (think of one subject in a room and one on the other end of the phone) and movement of a sound source or microphone. A further cause which pertains only to background sounds is that of ANC (active noise cancellation) used by mobile phones, which attenuates the magnitude of the background sounds (Brixen, 2019) to aid in improving speaker intelligibility.

Effect

It is common for forensic recordings to be relatively low in volume, particularly when captured with a portable device. This can result in only a fraction of the available quantisation levels being used, specifically at the LSB (least significant bit), and subsequently, a coarse representation of the acoustic environment captured. Care should be taken as to the placement of compression within the chain as it is at the LSB level where the quantisation error is most prevalent, so increasing the level of the signal too early in the chain can result in unwanted artefacts becoming audible. By applying selective forms of gain in a considered order, the extent of quantisation levels occupied can be increased, thus improving the desired signal level while limiting artefacts.

The desired signal level will also be closer to the noise floor within quiet recordings, so audible quantisation noise and masking can become an issue. Invariably high amplitude recordings suffer from the opposite fate, in that they may result in clipping and distortion.

The local effects of variations in amplitude are perceived as magnitude disparities between regions. At its most extreme, there may be a single voice of high amplitude, and a second which is barely audible due to a low-level capture. Decreasing the magnitude of the louder voice will remove data relating to this voice, so is not ideal for forensics work, and increasing the amplitude of the lower level voice will result in increased audibility of noise, as the signal to noise ratio is set at capture. It is a dilemma which requires compromise.

Solution

Gain refers to the process by which a post-processed signal will 'gain' a certain level of power with regards to the input (although it can also apply to attenuation if a negative value is used) and is most commonly measured in dB.

There are various methods available to implement this enhancement, and changes can be made locally or globally based on pre-defined settings by the user. These can result in manual or automated gain dependent on the method and settings applied. The various options available will now be detailed.

COMPRESSION

Through the careful use of dynamic range compression, it is possible to increase the mean signal level while maintaining the peak signal level. This allows lower magnitude signals that may be present, such as voices and background noises of interest to be audible, but in doing so increases the SNR by bringing the level of the noise floor upwards. It can also change the dynamics of the signal, which could render the speech signal of lower quality. This would likely reduce the accuracy of further tasks which may follow, such as voice comparisons or gunshot analysis.

In terms of automated methods, the most effective is that of an upwards expander (so named as it performs the opposite function to a compressor), which will provide amplification to low-level regions while ignoring those of a higher level. A downward compressor will give the same result but through the attenuation of higher levels relative to those that are lower. Typical characteristics of a compressor are the range, ratio, and envelope settings. The range determines the level at which the compressor acts upon the signal. The ratio is the amount of gain reduction applied to the input, and the envelope settings allow the time it takes for the compressor to act (attack) and stop (release) to be adjusted. Fast attack times can smooth out increases in signal level, and longer release times are required to prevent excessive distortion of the speech envelope (Hoare et al., 2008). Care should be taken with all settings to avoid artefacts of pumping caused by perceptual changes to the signal level caused by the compression or expansion.

MANUAL GAIN

A more time-consuming but precise method is that in which the examiner manually increases or decreases the gain for specific regions of the recording based on whether there is an area in which the level is too high or low relative to the entire recording. This is often the case in forensic audio, especially when recordings involve multiple speakers at differing distances from the microphone. For example, in the case of an informant wearing a wire, the voice of the individual wearing the microphone will always have the highest signal level as they are within closest proximity of the diaphragm, so the suspect's voice will need amplifying or the informant's voice will need reducing. In a studio mixing environment, it is often encouraged to reduce the level of the loudest signal rather than increase the quiet levels as this brings up the noise floor. As reducing the level will result in a loss of information from the recording, it makes more sense in forensic audio to bring the lower levels up to maintain as much of the original signal as possible. Applying some light compression once this has been performed can smooth out any level changes, analogous to smoothing the icing on the top of a cake to remove any slight bumps or ripples. Manual gain changes are the recommended technique as they allow the greatest degree of control, but time constraints can render it impractical for recordings of extended durations.

LIMITERS

Limiters are compressors which operate at a ratio of around 10:1 or above. At this ratio, the signal which passes over the threshold is being attenuated at such a high ratio (for every 10 dB over the threshold the output is attenuated by 9 dB) that it is limiting the signal. Limiters are useful in situations when there may be short bursts of high-level sound among a relatively low-level recording, such as gunshots within a quiet suburban environment (Izhaki, 2008).

NORMALISATION

Once disparities between the levels of different regions have been resolved, a final global gain adjustment may be required to optimise the playback level. This is achieved through the process of normalisation.

It is accepted that audio at a low signal level is difficult to hear, so the user will more often than not turn up the volume control of the sound interface (and thus the headphones or speaker), but this may still result in a less than optimal signal, and is far from ideal, especially when it may be played across a variety of systems once provided to the instructing party. If the gain is applied manually, the final level may not be as high as it could be, or result in clipping, causing loss of information and potential distortion. The solution is normalisation.

Normalisation can apply to both amplitude and frequency, but for the purpose of audio enhancement, it is applied to amplitude. It is the process of increasing the signal level as much as possible without introducing clipping and the associated quality degradation. For instance, if the maximum peak signal of a recording is -20 dBFS, the signal can be increased by up to 20 dB (although in reality a small amount of headroom is advised to prevent inter-sample clipping). Rather than applying gain in various areas of the signal, such as a compressor does, the increase in level is applied to the entire signal indiscriminately (Katz, 2002). The issue with this process is that quantisation noise will increase as the recording has already been quantised, especially on recordings of 16 bits or less. Increasing the level will leave the original SNR unchanged and add more even quantisation noise as the audio is being re-quantised, rendering the recording noisier than before. If normalisation is applied at the beginning of the enhancement process, the headroom will be reduced, which may be required for any processing which follows.

There are several distinct methods for normalisation, based on either the signal peaks (peak volume detection) or the overall level of a signal (RMS volume detection). Peak volume detection only considers the peaks of the waveform and is used when aiming to achieve the loudest volume possible from the signal (International Telecommunications Union, 2015). It is the safest approach as it can be ensured that the amplitude peaks within the signal will not exceed the normalisation settings applied by the user, resulting in clipping. RMS volume detection calculates the RMS of the signal, which is mathematically represented as (Brixen, 2017):

RMS = voltage peak x 0.707

RMS does not consider the equal loudness contours of the human ear, which could result in frequencies that are insensitive to the human auditory system being given more weight than is necessary, causing unnatural level changes. Due to these issues, a final method is available, EBU R-128 Volume Detection (European Broadcasting Union, 2010), which does take into account the perception of sound by the human ear, and so normalised recordings are perceived to be more consistent in their volume. When using either the RMS or EBU methods, care should be taken to ensure peaks do not exceed 0 dBFS, thus preventing clipping. If the previous processing stages have been carried out correctly, then all normalisation methods available are acceptable, providing the limitations of each are taken into account, and the waveform is visually analysed post-processing to confirm no clipping has occurred.

A final consideration is inter-sample peaks. These are peaks that occur between the quantised sample values, and as such, are not accounted for when peak levels are measured. If a signal is close to 0 dBFS, upon reaching the DAC, the signal can clip, depending on the playback equipment used (Nielsen and Lund, 2000). Although it can be argued that inter-sample peaks are unnoticeable, they can, in theory, cause small distortions due to clipping, so it is best to leave some headroom when normalising to prevent this occurrence. In practice, a -1.00 dBFS peak value is a good compromise to avoid clipping while achieving a strong signal level.

Order of operations

All enhancements are a culmination of a series of processes, and as such, each process interacts with the next in the chain. Although there are processes which can be performed at any point during the enhancement, there are some which should be performed before others in order to achieve the optimal processing from each tool, and thus, an optimal enhancement.

The following provides a proposed framework and the reasons for the position of each stage within the sequence. In the majority of cases, only a few of the processes will be required, but the general order should be maintained regardless of the amount of processing applied.

The key to ensuring the best result possible requires the optimal enhancement at every stage of the sequence by removing superfluous elements where possible. An example would be reducing the bandwidth of a signal to increase the efficacy of the noise reduction process which follows. Each enhancement process should be carefully considered and applied only when necessary, as the use of extraneous techniques will increase quantisation noise and potentially generate a final recording that is of lower quality and intelligibility than the original.

Step 1: Fix distortions

Transient noises are unsuitable for processing by complex algorithms, such as adaptive filters and spectral subtraction due to their unpredictable nature. If these are not removed before processing, they can cause the algorithms to react in ways which reduce their overall efficacy by using singular events such as random clicks to predict the behaviour of future samples. All transient distortions should, therefore, be removed as the first step of an enhancement. When fixing distortions, the region being processed must be accurately selected to prevent the unnecessary removal of surrounding information. Magnification of the waveform and careful positioning of the start and endpoints will ensure this is achieved.

Step 2: Stereo source separation

Source separation can be the first stage of processing, providing there are no distortions to be fixed. It could also be ignored if not applicable, such as if the recording consists of a single channel without any background music. As reference cancellation relies on the information within the recordings to be as closely matched to the original as possible to be efficient, any processing larger than the removal of transients will change samples within the evidence recording that are required for efficient processing using the reference signal. Consider that reference cancellation requires the evidence signal to match the reference signal as closely as possible. If the relationships between samples are significantly changed within the evidence signal, it will render the process impossible. In terms of the amount of redundant data removed, source separation removes the largest amount, and as such, this processing step should be performed as soon as possible in the enhancement processing chain. It also allows the stages which follow to focus on the desired signal and the remaining noise surrounding it.

Step 3: Remove static redundant data

As the desired signal is often limited to a specific bandwidth relative to the entire signal spectrum, the surrounding, redundant data can be removed through the use of filters. For example, if speech is the desired signal, the frequencies below 100 Hz may be removed to provide further clarity to the desired area, analogous to removing the weeds to expose the flowers. Performing this procedure before complex adaptive processes such as de-reverberation and dynamic noise reduction saves on processing power, allowing the algorithms to be more flexible and optimised by focusing on only the desired area. There is little sense in performing complex processing techniques on any elements of the recording which are extraneous to the desired signal.

Step 4: Gain correction stage one

As many algorithms process future samples based on previous samples, any high-level signal bursts will cause the processor to react in ways that reduce its efficiency. To counteract this, any static momentary signal bursts of a prominent level relative to the rest of the signal should be attenuated through manual gain reduction, compression, or limiting. It is also vital that levels of the desired signal are matched if there is a large level difference between them. This allows optimising of the thresholds for future processing, preventing the lower level signal being misinterpreted as noise and removed by noise removal techniques.

Step 5: Remove dynamic redundant data

Once noise which may reduce both the processing power and efficacy of complex processors has been removed, the focus can shift to getting rid of the remaining redundant data, the location of which can be found through FFT analysis and critical listening. Processors which remove dynamic data are determined by machine learning, so by exhausting all manual processes available, it can be ensured that they function at their optimum level. Care should always be taken, to ensure information that is part of the desired signal is not lost. This can be achieved by listening to the audio that is being rejected by the process if the enhancement software allows. This step is performed in order to remove the noise that the static removal stage could not. It occurs after the first gain correction stage as it does require a degree of machine learning, which may become confused if there are high-level signal bursts and low SNR signal areas within the signal.

Step 6: Remove noise

Automatic techniques for noise removal can now be employed as all manual methods have been exhausted. This stage must be placed before de-reverberation as reverb involves a specific convolution which smudges transients, and any noise which is masking these areas will confuse the de-reverberation algorithm, rendering it less effective. If de-reverberation is performed before noise removal, the algorithm may struggle to differentiate between noise and reverberation and will, therefore, be less effective than it could be in completing the task. For optimal performance, a region with a similar noise profile to the area in which the de-noising is being applied should be selected during the learning stage. If there are changes to the environment during the recording, for example, indoor to outdoor, two specific noise removal instances will most likely be required. It is not uncommon for multiple passes of de-noising at different settings to be performed in order to achieve optimal results.

Step 7: De-reverberation

The removal of reverberation should be applied once all other noise has been removed, as the signal transients and indirect reflections pertaining to the desired signal should now be exposed. Reverberation is highly correlated to the signal, so any noise surrounding it will cause erroneous processing which could result in both increased noise and less than optimal results.

Step 8: Gain correction stage two

A final gain correction is employed at this stage to ensure a comparable magnitude of the desired signal throughout the recording. Manual level changes or an expander to match levels of two speakers can be considered, as the final levels can now be judged accurately. Careful adjustments to the frequency spectrum can also be made to enhance clarity and listenability, such as adding brightness through shelving filters, or attenuation of lower frequencies to reduce the proximity effect.

Step 9: Normalisation

The final stage is to ensure the overall magnitude of the audio is at a level which guarantees listenability on all systems. Performing this before any processing would apply an additional and unnecessary round of quantisation noise and reduce signal headroom, leaving less room for processing, hence its position at the end of the enhancement process chain. When performed on the final signal, the increase in noise is minimised as it has previously been removed by all other processes within the framework. It is recommended to normalise to -1 dBFS using the peak detection algorithm for optimal amplitude without clipping.

Step 10: Time-stretching

On occasions when the intelligibility of speech is paramount, time-stretching may be required as it can aid in improving the intelligibility of speech, especially when the cadence is fast. The speech should never be reduced below 75 per cent of the original speed, and the pitch should be maintained to prevent misinterpretation. This is performed after all other stages, as although it is included in the enhancement sequence, the desired signal is not being enhanced in terms of quality. To put this stage anywhere other than the final step, therefore, would make no sense. A reduced playback speed version should be provided in addition to, not instead of the original playback speed version.

Step 11: File output

The final stage of an audio enhancement is to output the generated exhibit in a non-proprietary format for playback on all standard media players, thus ensuring it will be playable for the instructing party and end user. The format should be in a standard uncompressed encoding format to guarantee data is not lost in the process. Caution should be used with regards to the sample rate selected as down-sampling from the original rate will result in a loss of information, so this should never be practised, and up-sampling can result in changes to the audio due to the interpolation of the samples. Exporting at the sample rate at which processing was performed is, therefore, the safest practice. As the audio is re-encoded when exported, using a high bit depth will ensure a high signal to noise ratio and thus mitigate against the audibility of quantisation noise. The hash checksum should be calculated, preferably using an algorithm that has no history of being compromised, as it has been shown that both SHA-1 (Stevens et al., 2017) and MD5 (Wang et al., 2004) have had collisions, meaning two files have been shown to have matching hash checksums. Although the level of effort to compromise these hashes is extreme, by using an uncompromised algorithm, it prevents any questions over the integrity of a file being raised.

Step 12: A/B comparison

Although comparisons should be made throughout the enhancement to determine the degree of success of each processing stage, a final comparison should always be made against the original version to ensure the processing has improved the recording, has not reduced intelligibility and has not introduced audible artefacts. As previously stated, although higher quality does not equate to increased intelligibility, it is rarely the case that a recording will become lower in quality but more intelligible. If the final A/B comparison results in the opinion that the enhancement was unsuccessful, steps should be taken to discover why and to remedy those issues.

Step 13: Peer review

The previous A/B comparison should finally be performed by another examiner to obtain a second opinion as to the success of the enhancement.

Enhancement reporting

Reports relating to an enhancement should contain similar content to that of other forensic reports, although obviously no opinion will be included. There are also a few specifics in relation to these reports which may not appear in others, such as the following.

Exhibit quality

The quality of the exhibit should be documented, in terms of the specifications, issues pertaining to the clarity of the audio, and potentially the intelligibility of speech. In doing so, the limitations caused by such are made clear.

Enhancement process

The process should be documented in terms of the sequence and settings of processes applied. As these should be documented in contemporaneous working notes, the report version does not need to be as detailed, as the intended audience of the report does not require technical information. It should be noted within the report that the exact settings used for each process can be made available on request to aid in transparency while evidencing that a methodology was followed.

Generated exhibits

The details concerning the enhanced audio exhibit should be documented to ensure a chain of custody can be started. Imagine a scenario where the instructing party decided to convert the enhancement to a lower quality format for transmission by email, and in the process, some important data was lost. If questions are raised in relation to this at a later date, the details of the original exhibit generated by the laboratory are readily available. The generated exhibit information should contain, at minimum, the filename, format, date generated, and hash checksum.

Transcriptions

When audio recordings are submitted as evidence, it is common for the submitting party to provide a transcript to mitigate against objections as to what was said within the recording later down the line. In many cases, the speech is intelligible, to the degree that the majority of people would create identical transcriptions. When there is a disagreement between the parties as to what is said, both will submit a transcription, and the jury will be provided with both.

One of the most common misconceptions by laypersons is that enhancement can improve the intelligibility of a recording. This is rarely the case, for the following reasons:

1   The speech data has not been captured to the degree that would allow it to be easily intelligible (for example, whispering).
2   A person is talking in a language other than their first, thus pronouncing words differently.
3   Unnatural speech (such as shouting).
4   Distorted speech (such as that which contains clipped samples) (Cerrato and Paoloni, 1999).

When any of the above is the case, there is no specific scientific methodology which can be performed to make the data required for a complete transcription appear, and as such, there is no 'expertise' in the transcription of disputed utterances. Phoneticians, however, have methodologies in the decoding of competing speakers, disguised voices, and speech influenced by stress, drugs, and alcohol.

Further to this, if the speech is already of poor quality due to an insufficient amount of data being captured in relation to such, the slightest enhancement may have a large impact. In fact, the less data captured of the speech, the higher the potential for misinterpretation. Processing could cause the word to become intelligible, but there is no way of knowing if it was the actual word spoken or whether it just sounds like a particular word because of changes made to the data through processing.

With that being said, audio forensic examiners have several advantages over triers of fact:

1   Understanding of bias so can mitigate against it.
    As with all forensic processes, biases must be limited as far as possible. Biasing information may include not only the case details, but also the context, the surrounding words, and other words in the recording that may influence the interpretation.
2   Specialist equipment
    Audio forensic examiners have access to high-end digital to analogue converters and flat frequency response headphones. In contrast, triers of fact will more often than not listen on low specification court computers or laptops through small speakers with a frequency response that colours the sound. When taking part in cases in which the audio is essential, and where possible (such as if the trier of fact is a single person), headphones and an external sound card should be provided to afford them the optimal listening experience. It is the duty of the examiner to ensure the audio is as clear as possible, so if the listening experience can be improved, efforts should be made to do so. The inclusion of advice on the playback of recordings can also be provided within a report appendix to aid the subject listening to the recording where it is not possible to be present during the listening session.
3   Controlled environment
    Forensic examiners work in quiet surroundings and have the opportunity to listen to recordings multiple times. People in a courtroom usually listen to the recording a couple of times at

Table 10.4 Presentation of transcriptions

Offset	Source	Speech/Action
00:03	MALE 1	What did you (think/drink)?
00:05	MALE 2	I can't remember (1–2 non-intelligible words)
00:06		*(ring tone in the background)*
00:09	(MALE 1/MALE 2)	Let's go and see Brian

Source: (Poza and Begault, 2008).

most, and as they are played through loudspeakers, there are also external noises and reflections off the surfaces of the room which will only reduce the intelligibility of speech.

The best approach for forensic examiners when providing transcriptions is to be transparent, and as with all forensic work, document the limitations, the nature of non-intelligible areas (in either the number of words, syllables, or duration), and use a scale of confidence for words transcribed. The latter can be easily implemented through the use of a key, such as:

word – full confidence in the word due to high intelligibility
(word) – less confidence in the word in brackets relative to the rest of the recording
(word1/word2) – less confidence and two words are suggested
(*word*) – action of some sort
SPEAKER – full confidence in the identity of the speaker
(SPEAKER) – less confidence in the identity of the speaker relative to the rest of the recording
(SPEAKER 1/SPEAKER 2) – less confidence in the identity of the speaker and two speakers are suggested

Timing relating to both the transcribed speech and the speaker should also be documented for clarity and ease of reference. An example is provided in Table 10.4.

Summary

The enhancement of audio is probably the area in which audio forensics is most commonly associated due to coverage in various films and popular television shows. It is also an area that requires no expertise for an opinion to be given as to its success as quality is a subjective measure. Expectations can, therefore, be high, and managing them by explaining and being transparent about the limitations is a critical aspect of the work. In order to do this and to ensure that audio is enhanced to the highest level of quality and intelligibility possible, all issues and remedies for these issues must be understood.

References

Aleinik, S. and Matveev, Y.N., 2014. Detection of clipped fragments in speech signals. World Academy of Science, Engineering and Technology 86, 703–709.

Alexander, A. and Forth, O., 2011. 'No, thank you, for the music': An application of audio fingerprinting and automatic music signal cancellation for forensic audio enhancement. Paper presented at 2011International Association of Forensic Phonetics and Acoustics Conference, Vienna, Austria.

Alexander, A., Forth, O., and Tunstall, D., 2012. Music and noise fingerprinting and reference cancellation applied to forensic audio enhancement. Paper presented at 46th AES International Conference: Audio Forensics. Audio Engineering Society.

Alves, R.G., Yen, K.-C., Vartanian, M.C., and Gadre, S.A., 2010. Method to improve speech intelligibility in different noise conditions. Paper presented at the 128th AES Convention, Audio Engineering Society, London.

American National Standards Institute, 1997. Methods for calculation of the speech intelligibility index. Available at: webstore.ansi.org/Standards/ASA/ANSIASAS31997R2017

ANSI/ASA, 2009. Method for measuring the intelligibility of speech over communication systems. Available at: webstore.ansi.org/Standards/ASA/ANSIASAS32009R2014

Betts, D., French, A., Hicks, C., and Reid, G., 2005. The role of adaptive filtering in audio surveillance. Paper presented at 26th AES International Conference: Audio Forensics in the Digital Age, Audio Engineering Society.

Betts, D., Reid, G., and Chan, D., 1992. Application of digital signal processing to audio restoration. Paper presented at UK 7th Conference: Digital Signal Processing (DSP), Audio Engineering Society, UK.

Bohn, D.A., 1988. Operator adjustable equalisers: An overview. Paper presented at the 6th International Conference: Sound Reinforcement, Audio Engineering Society, Nashville, USA.

Boll, S. F. 1979. Suppression of acoustic noise in speech using spectral subtraction. IEEE Transactions on Acoustics, Speech and Signal Processing 27(2).

Bolt, R.H. and MacDonald, A.D., 1949. Theory of speech masking by reverberation. The Journal of the Acoustical Society of America 21, 577–580.

Brandt, M. and Bitzer, J., 2012. Hum removal filters: Overview and analysis. Paper presented at the 132nd AES Convention, Audio Engineering Society, Budapest, Hungary.

Brandt, M. and Bitzer, J., 2014. Automatic detection of hum in audio signals. Journal of the Audio Engineering Society 62, 584–595.

Brixen, E.B., 2017. Audio metering: Measurements, standards and practice, 2nd edn. Routledge, London.

Brixen, E.B., 2019. Directivity and sensitivity of cell phones: iPhone 7. Paper presented at the International Conference on Audio Forensics, Audio Engineering Society, Porto, Portugal.Brixen, E.B. and Hensen, R., 2006. Wind generated noise in microphones: An overview, Part 1. Paper presented at the 120th AES Convention, Audio Engineering Society, Paris, France.

Cerrato, L. and Paoloni, A., 1999. Are transcriptions of speech material recorded by means of bugs reliable? Paper presented at the Sixth European Conference on Speech Communication and Technology, Budapest, Hungary.

Chalil, M., Raghotham, D., and Dominic, P., 2008. Smooth PCM clipping for audio. Paper presented at the 34th International Conference, Audio Engineering Society, Jeju Island, Korea.

Cole, D.R., Moody, M.P., and Sridharan, S., 1995. Robust enhancement of reverberant speech. Paper presented at the 5th Australian Regional Convention, Audio Engineering Society, Sydney, Australia.

Datta, J., Zhou, X., Begin, J., and Martin, M., 2018. An auditory model-inspired objective speech intelligibility estimate for audio systems. Paper presented at the 144th AES Convention, Audio Engineering Society, Milan, Italy.

Desmond, J.M., Collins, L.M., and Throckmorton, C.S., 2014. The effects of reverberant self- and overlap-masking on speech recognition in cochlear implant listeners. The Journal of the Acoustical Society of America 135, 304–310.

Dobre, R.A., Nita, V.A., Ciobanu, A., Negrescu, C., and Stanomir, D., 2015. A hum removal algorithm used for audio restoration purposes. Paper presented at the 2015 International Symposium on Signals, Circuits and Systems (ISSCS), IEEE, Iasi, Romania.

Ellis, D., 2009. Robust landmark-based audio fingerprinting. Available at: www.ee.columbia.edu/~dpwe/resources/matlab/fingerprint

European Broadcasting Union, 2010. R 128: Loudness normalisation and permitted maximum level of audio signals. Available at: tech.ebu.ch/publications/r128

Fairbanks, G., 1958. Test of phonemic differentiation: The rhyme test. The Journal of the Acoustical Society of America 30, 596–600.

Fechner, N., Kirchner, M., 2014. The humming hum: Background noise as a carrier of ENF artifacts in mobile device audio recordings. Paper presented at the 2014 Eighth International Conference on IT Security Incident Management & IT Forensics (IMF), IEEE, Münster, Germany.

French, N.R., Steinberg, J.C., 1947. Factors governing the intelligibility of speech sounds. Journal of the Acoustical Society of America 19.

Gaspar, I., 1992. Practical considerations on overload-distortion of analog-to-digital converters. Paper presented at the 92nd AES Convention, Audio Engineering Society, Vienna, Austria.

Gaubitch, N.D., Thomas, M.R.P., and Naylor, P.A., 2010. Dereverberation using LPC-based approaches. In P.A. Naylor and N.D. Gaubitch (eds), Speech dereverberation. Springer, London, pp. 95–128. Available at: https://doi.org/10.1007/978-1-84996-056-4_4

Godsill, S.J. and Rayner, P.J.W., 1998. Digital audio restoration: A statistical model based approach. Springer, London.

Goodman, D., Lockhart, G., Wasem, O., and Wong, W-C., 1986. Waveform substitution techniques for recovering missing speech segments in packet voice communications. IEEE Transactions on Acoustics, Speech, and Signal Processing, 34, 1440–1448.

Gray, R.M. and Davisson, L.D., 2004. An introduction to statistical signal processing. Cambridge University Press, Cambridge.

Grigoras, C., 2009. Applications of ENF analysis in forensic authentication of digital audio and video recordings. Journal of the Audio Engineering Society 57, 643–661.

Habets, E.A.P., 2010. Speech dereverberation using statistical reverberation models. In P.A. Naylor and N.D. Gaubitch (eds), Speech dereverberation. Springer, London, pp. 57–93.

Hansen, J.H.L. and Pellom, B.L., 1998. An effective quality evaluation protocol for speech enhancement algorithms. Paper presented at the 5th International Conference on Spoken Language Processing, Sydney, Australia.

Hicks, C. and Reid, G., 1996. Evolution of broadband noise reduction techniques. Paper presented at the 6th Australian Regional Convention, Audio Engineering Society, Melbourne, Australia.

Hoare, S., Hughes, P., and Turnbull, R., 2008. Audio enhancement for portable device based speech applications. Paper presented at 124th AES Convention Audio Engineering Society, Amsterdam, Netherlands.

House, A.S., Williams, C.E., Hecker, M.H.L., and Kryter, K.D., 1965. Articulation-testing methods: Consonantal differentiation with a closed-response set. The Journal of the Acoustical Society of America 37, 158–166.

Hu, Y. and Loizou, P.C., 2007. A comparative intelligibility study of single-microphone noise reduction algorithms. The Journal of the Acoustical Society of America 122, 1777–1786.

Hu, Y. and Loizou, P.C., 2008. Evaluation of objective quality measures for speech enhancement. IEEE Transactions on Acoustics, Speech, and Language Processing 16, 229–238.

Huber, D.M. and Runstein, R.E., 2005. Modern recording techniques, 6th edn. Elsevier, Oxford.

Huckvale, M. and Frasi, D., 2010. Measuring the effect of noise reduction on listening effort. Available at: www.aes.org/e-lib/browse.cfm?elib=15501

Ignatov, P., Stolbov, M., and Aleinik, S., 2012. Semi-automated technique for noisy recording enhancement using an independent reference recording. Paper presented at 46th International Conference: Audio Forensics, Audio Engineering Society, Denver, CO, USA.

International Organization for Standardization, 1996. P.800 Methods for objective and subjective assessment of quality. Available at: global.ihs.com/doc_detail.cfm?item_s_key=00255796

International Organization for Standardization, 2001. ISO 9921:2003. Available at: www.iso.org/obp/ui/ #!iso:std:33589:en

International Telecommunications Union, 2015. BS.1770: Algorithms to measure audio programme loudness and true-peak audio level. Available at: www.itu.int/dms_pubrec/itu-r/rec/bs/R-REC-BS. 1770-0-200607-S!!PDF-E.pdf

Izhaki, R., 2008. Mixing audio: Concepts, practices and tools. Elsevier Ltd, Oxford.

iZotope, 2014. How to use dynamic EQ in mastering. Available at: www.izotope.com/en/learn/how-to-use-dynamic-eq-in-mastering.html

Janssen, A., Veldhuis, R., and Vries, L., 1986. Adaptive interpolation of discrete-time signals that can be modeled as autoregressive processes. IEEE Transactions on Acoustics, Speech, and Signal Processing 34, 317–330.

Kamath, S. and Loizou, P., 2002. A multi-band spectral subtraction method for enhancing speech corrupted by colored noise. In Proceedings of 4 ICASSP. Orlando, Florida, USA. pp. 44164–44164.

Katz, B., 2002. Mastering audio: The art and the science. Focal Press, Waltham, MA.

Kinoshita, K., Nakatani, T., Miyoshi, M., and Kubota, T., 2008. A new audio post-production tool for speech dereverberation. Paper presented at the 125th AES Convention, Audio Engineering Society, San Francisco, USA.

Kodrasi, I., Cauchi, B., Goetze, S., and Doclo, S., 2017. Instrumental and perceptual evaluation of dereverberation techniques based on robust acoustic multichannel equalisation. Journal of the Audio Engineering Society 65, 117–129.

Koenig, B.E., Lacey, D.S., and Killion, S.A., 2007. Forensic enhancement of digital audio recordings. Journal of the Audio Engineering Society 55, 352–371.

Kraght, P., 2000. Aliasing in digital clippers and compressors. Journal of the Audio Engineering Society 48, 1060–1065.

Kryter, K.D., 1962. Methods for the calculation and use of the articulation index. The Journal of the Acoustical Society of America 34, 1689–1697.

Lagrange, M. and Marchand, S., 2005. Long interpolation of audio signals: Using linear prediction in sinusoidal modeling. Journal of the Audio Engineering Society 53, 891–905.

Laguna, C. and Lerch, A., 2016. An efficient algorithm for clipping. Paper presented at the 141st AES Convention, Audio Engineering Society, Los Angeles, USA.

Lim, J., 1978. Evaluation of a correlation subtraction method for enhancing speech degraded by additive white noise. IEEE Transactions on Acoustics, Speech, and Signal Processing 26, 471–472.

Lorber, M. and Hoeldrich, R., 1997. A combined approach for broadband noise reduction. Paper presented at the 1997 Workshop on Applications of Signal Processing to Audio and Acoustics, IEEE, New Paltz, NY, USA.

Łuczyński, M., 2019. Primary study on removing mains hum from recordings by active tone cancellation algorithms. Paper presented at the 146th AES International Convention, Audio Engineering Society, Dublin, Ireland.

Luknowsky, D. and Boyczuk, J., 2004. Audio processing in police investigations. Canadian Acoustics 32, 154–155.

Ma, J., Hu, Y., and Loizou, P.C., 2009. Objective measures for predicting speech intelligibility in noisy conditions based on new band-importance functions. Journal of Acoustic Society of America 125, 3387.

Maher, R.C., 2005. Audio enhancement using nonlinear time-frequency filtering. Paper presented at 26th AES International Conference: Audio Forensics in the Digital Age, Audio Engineering Society.

Mahieux, Y. and Marro, C., 1996. Comparison of dereverberation techniques for videoconferencing applications. Paper presented at the 100th Audio Engineering Society Convention, Audio Engineering Society, Copenhagen, Denmark.

Mapp, P., 2016. Intelligibility of cinema & TV sound dialogue. Paper presented at 141st Audio Engineering Society Convention, Audio Engineering Society, Los Angeles, USA.

Mathai, M.K. and Deepa, J., 2015. Design and implementation of restoration techniques for audio denoising applications. Paper presented at the 2015 IEEE Recent Advances in Intelligent Computational Systems (RAICS), IEEE, Trivandrum, Kerala, India.

Naruka, K. and Sahu, O.P., 2015. Objective quality and intelligibility evaluation for speech enhancement algorithms. Journal of Basic and Applied Engineering Research 2, 1891–1895.

Nees, S., Schwarz, A., and Kellermann, W., 2016. Dereverberation using a model for the spatial coherence of decaying reverberant sound fields in rectangular rooms. Paper presented at the 60th AES International Conference, Audio Engineering Society, Leuven, Belgium.

Nielsen, S.H. and Lund, T., 2000. 0dBFS+ levels in digital mastering. Paper presented at the 109th AES Convention, Audio Engineering Society, Los Angeles, USA.

Nielsen, S.H. and Lund, T., 2003. Overload in signal conversion. Paper presented at the 23rd International Conference, Audio Engineering Society, Copenhagen, Denmark.

Oppenheim, A.V., Schafer, R.W., and Stockham, T.G., 1968. Nonlinear filtering of multiplied and convolved signals. IEEE Transactions on Audio and Electroacoustics 16.

Poza, F. and Begault, D.R., 2008. The role of transcriptions in the courtroom: a scientific evaluation. Paper presented at the 33rd AES International Conference, Audio Engineering Society, Denver, CO.

Ramirez, J.L., 2016. Effects of clipping distortion on an automatic speaker recognition system. National Center for Media Forensics, University of Colorado, Denver, CO,

Riemer, T.E., Weiss, M.S., and Losh, M.W., 1990. Discrete clipping detection by use of a signal matched exponentially weighted differentiator. Paper presented at the Southeastcon 90, IEEE.

Rix, A.W., Beerends, J.G., Hollier, M.P., and Hekstra, A.P., 2001. Perceptual evaluation of speech quality (PESQ) – a new method for speech quality assessment of telephone networks and codecs, Paper presented at the 2001 IEEE International Conference on Acoustics, Speech, and Signal Processing. Proceedings, IEEE, Salt Lake City, UT, USA. Available at: https://doi.org/10.1109/ICASSP.2001.941023

Rix, A.W., Hollier, M.P., Beerends, J.G., and Hekstra, A.P., 2000. PESQ: the new ITU standard for end-to-end speech quality assessment. Paper presented at 109th AES Convention, Audio Engineering Society.

Sadjadi, S.O. and Hansen, J.H.L., 2012. Blind reverberation mitigation for robust speaker identification, Paper presented at the ICASSP 2012, 2012 IEEE International Conference on Acoustics, Speech and Signal Processing, IEEE, Kyoto, Japan.

Shujau, M., Ritz, C.H., and Burnett, I.S., 2013. Speech dereverberation based on linear prediction: An acoustic vector sensor approach. Paper presented at the ICASSP 2013, 2013 IEEE International Conference on Acoustics, Speech and Signal Processing (ICASSP), IEEE, Vancouver, BC, Canada.

Singh, L. and Sridharan, S., 1998. Speech enhancement using critical band spectral subtraction. Paper presented at International Conference on Speech and Language Processing, Sydney, Australia.

Sørensen, T.B., 2005. Click and pop measurement technique. Paper presented at the 27th International Conference, Audio Engineering Society, Copenhagen, Denmark.

Soulodre, G.A., 2010. About this dereverberation business: A method for extracting reverberation from audio signals. Paper presented at the 129th AES Convention, Audio Engineering Society, San Francisco, USA.

Steeneken, H.J.M. and Houtgast, T., 1980. A physical method for measuring speech-transmission quality. The Journal of the Acoustical Society of America 67, 318–326.

Stevens, M., Bursztein, E., Karpman, P., Albertini, A., and Markov, Y., 2017. The first collision for full SHA-1. In J. Katz and H. Shacham (eds), Advances in cryptology – CRYPTO 2017. Springer International Publishing, Cham, pp. 570–596.

Stolbov, M. and Ignatov, P., 2012. Speech enhancement technique for low SNR recording. Paper presented at 46th AES International Conference: Audio Forensics, Audio Engineering Society, Denver, CO, USA.

Tan, C.-T. and Moore, B.C.J., 2003. The effect of nonlinear distortion on the perceived quality of music and speech signals. Journal of the Audio Engineering Society 51, 1012–1031.

Telecommunication Standardization Sector of ITU, 2001. Perceptual Evaluation of Speech Quality (PESQ): An objective method for end-to-end speech quality assessment of narrow-band telephone networks for speech codecs. Available at: www.itu.int/rec/T-REC-P.862-200102-I/en

Veen, B.V. and Buckley, K., 1988. Beamforming: A versatile approach to spatial filtering. IEEE ASSP Magazine 5, 4–24.

Wang, Z., Feng, D., Lai, Z., and Yu, H., 2004. Collisions for hash functions MD4, MD5, HAVAL-128 and RIPEMD. IACR Cryptology ePrint Archive.

Chapter 11

Audio authentication

Introduction

As digital audio forensics is a discipline of forensic science, the final destination for examination results is the justice system, and as such, any conclusions drawn from an audio recording can have a real impact on the lives of all parties involved. It is therefore of the utmost importance that the source of a recording can be trusted, as covered in Chapter 3 on forensic principles. In the age of free and low-cost audio editing software, relatively few resources are required to manipulate an audio recording, although both the extent of editing performed and the ease at which it can be detected will generally be in direct correlation with the skill level and knowledge of the individual performing the edit. For example, most smartphone audio recording applications now contain a 'trim' function to allow a user to remove sections from the beginning or the end of a recording. The reasons for such are many, from innocently removing redundant regions to meet email attachment limitations, or nefariously removing a region that somebody does not want to be heard. If a person is capable of making a recording with one of these software applications, it could be argued that they are just as capable of trimming the recording. At the extreme end of audio manipulation are voice synthesis techniques which build a model of a human voice using recordings of a real voice. A text-to-speech software program then allows a user to determine what the voice will say, whether it is a comment purporting to be by a co-worker to have them dismissed from their job or the confession to a crime by a co-accused. Somewhere in between the simple and the complex are several other manipulation techniques such as splicing, cloning, and deletion. It should be evident that any of the above would put justice at risk if they were deemed admissible, so any recordings which have been subject to such manipulation must be detected before they enter the justice system.

'Authentication' is the term for the process in which edited recordings are detected, as the examination is performed to 'authenticate' the recording, or provide an opinion concerning its 'authenticity.' Once an analysis has been completed, the consistency of the exhibit relative to an original is reported, as by definition, an original recording is not an edited recording. The Audio Engineering Society (2012) defines an authentic recording as one which is:

> made simultaneously with the acoustic events it purports to have recorded, made in a manner fully and completely consistent with the methods of recording claimed by the party who

produced the recording and free from unexplained artefacts, alterations, additions, deletions or edits.

It is defined by SWGDE (2016) as: 'the first manifestation of sound in a recoverable stored format be it a magnetic tape, digital device, voicemail file stored on a server, optimal disk, or some other form.'

Although this would technically require the purported original recording to perform authentication, things get cloudy with regards to bit-stream copies. A bit-stream copy is an exact digital copy of the original, and as such, every binary digit of which the original is composed is precisely the same as the copy. It is, therefore, acceptable to perform an authentication examination on a bit-stream copy of the original recording and is, in fact, best practice to mitigate against the risk of altering the original in any way.

As this concept can be difficult to understand for laypersons, the following are short scenarios for the reader to opine on whether they are original recordings or not.

1   An individual captures an MP3 recording, using their laptop, saves the file and then burns it to a playable audio CD to be provided as evidence.
2   An individual takes a WAV PCM recording on their Dictaphone, trims it using a built-in function to remove areas at the beginning and end which they deem unnecessary, takes the memory card out, connects it to their computer and moves the recording to their computer. They then put the recording onto a USB stick to be provided as evidence.
3   An individual makes an M4A recording using their mobile phone and sends the recording via text message to a friend. Their friend has the same mobile phone and downloads the recording onto their phone. The individual who made the recording deletes the recording from their phone and a week later the friend provides their phone as evidence.

The first scenario is probably the most common, and as the recording has been transcoded to meet the specifications of an audio CD (WAV PCM, 44.1 kHz, 16 bits), it is no longer a bit-stream copy of the original.

The second is also relatively common, and although the recording has not been transcoded to a new format, it does not represent the original recording as data has been removed post-capture.

The recording from scenario three may appear to be original at face value. Still, as it may have been transcoded during transfer, and some of the metadata may have changed to represent the time it was downloaded to the phone rather than the time it was initially captured, it is not an original version.

All of the above are real examples taken from casework and are provided to show the importance of attempting to obtain the most original recording rather than just accepting what has been submitted. Clear communication with the instructing party is required to help them understand what constitutes a bit-stream copy of the original recording.

An authentication (or at minimum a triage) should always be the first step of any examination, as to perform an enhancement on an edited recording only for it to be ruled inadmissible when it is served as evidence would consume both time and money. It may also mask the fact that it has been edited, as the now enhanced version is accepted as having been processed, so traces in relation to the edited version may be removed. For this reason, authentication is the most crucial step with regard to any audio exhibit, and a triage approach is essential before any further work is conducted.

Study of manipulations

To be effective in detecting manipulations, the possible edits which can be performed must first be understood. The following is, therefore, a short survey of the most common audio editing techniques.

Trimming/cropping

To trim an audio file is to remove data from the beginning or end of the recording, essentially 'trimming' it to a shorter length (Figure 11.1). This may be used nefariously to change the semantic context of a recording, or innocently to focus on only the pertinent section or to reduce the file size.

Deletion

Deletion pertains to the removal of audio data between the first and last sample of the recording, for which there are no innocent explanations. The area deleted could range from a single word to an extended region of the audio. Trimming is essentially a form of deletion, but the name 'trim' indicates that the data removal occurred at the extremes rather than within the duration of the content (Figure 11.2).

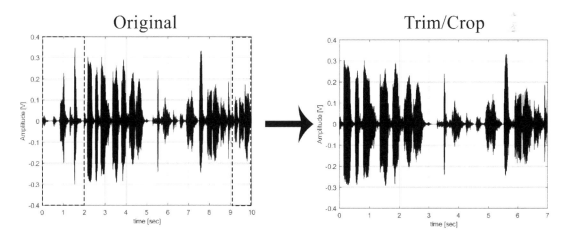

Figure 11.1 Trim function in which the regions within the broken lines are removed from a recording

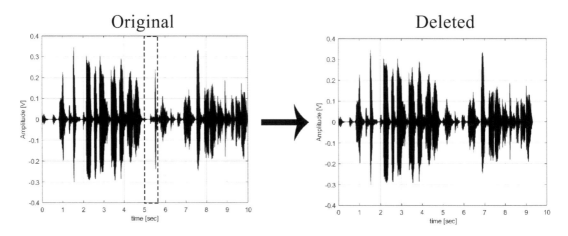

Figure 11.2 Delete function in which the region within the broken lines is removed from a recording

Splicing

There are various alterations which are subsets of splicing, but the essential principle is that audio has been added to the recording. This could be between the first and last sample of the recording, or appended to the beginning or end. It may be through cloning a section from one recording and splicing it into a new recording (Figures 11.3 and 11.4), cloning a section from within the same recording and splicing it to another area (Figures 11.5 and 11.6), or recording over a section within a recording. Again, there is no plausible innocent explanation for any form of splicing.

Speech synthesis

The synthesis of speech comparable to a human voice is a relatively new technique that has developed thanks to the progress made in the deep learning field (Galou and Chollet, 2011; Google Deepmind, 2016; Kumar et al., 2017; Sotelo et al., 2017; Starkie et al., 2002). The process involves taking recordings of an individual's voice and building a statistical model based on various elements of the speech. Once the model has been composed, a user can use a text to speech interface to program the voice to say whatever they desire. This method is still relatively novel, so both the complexity in creating a synthesised recording and the amount of data required are limiting factors, but it is expected that this area will become more prominent in the future.

It may be that only one of the above techniques is used, or all of them within a single recording. The good news for audio forensic examiners is that the more editing that is performed, the more traces that are left. With that being said, there are also steps which may be undertaken in an attempt to evade detection, so examiners should, therefore, do everything in their power to ensure both exhaustive and thorough examinations are performed.

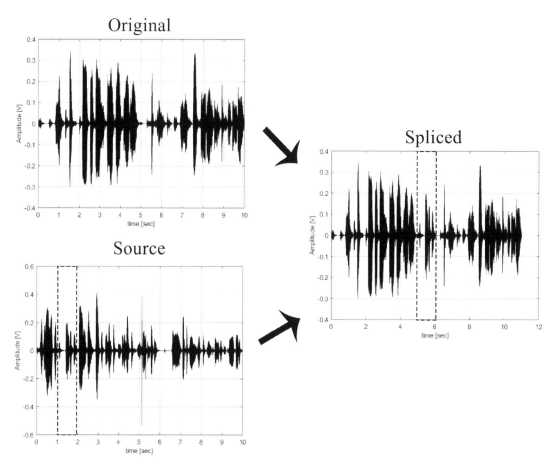

Figure 11.3 Splice function in which the region within the broken lines is inserted into an original signal, thus extending the duration by 1 second

Preparation

In preparation for an audio authentication examination, the exhibit recording must first be imaged from the purported capture device or medium on which it has been provided, and a working copy created.

As the aim of the examination is to determine whether the exhibit is consistent with an original recording, any information in relation to the purported capture device and conditions will assist in allowing comparisons to be made against such. This may appear contradictory to other advice with regards to bias mitigation, in which efforts are made to limit the information provided, but as authentication examinations require the testing of propositions, all information concerning the purported capture is useful. In many cases, there will be no obvious traces of the device within the

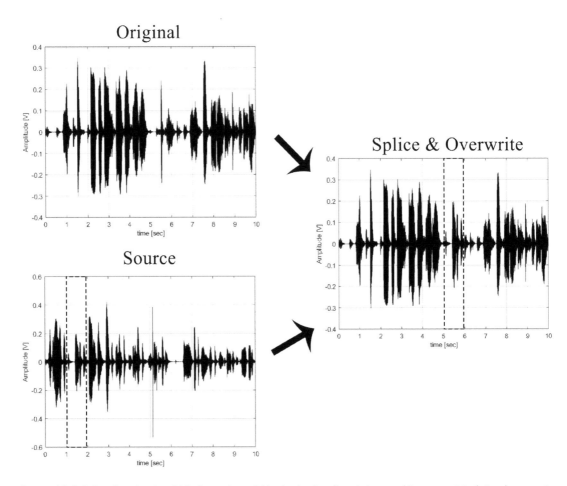

Figure 11.4 Splice function in which the region within the broken lines is inserted into an original signal, overwriting an area within. The duration, therefore, does not increase.

recording, so obtaining that information allows an opinion to be based on other findings. To that end, the instructing party should be provided within an authentication questionnaire (Cooper, 2006). This consists of various questions which can aid the examination, covering topics such as, but not limited to:

1 Details of the device with which the recording is purported to have been captured, including model, make, firmware, and operating system version.
2 Details of software used.
3 The environment in which the recording was captured.
4 The location in which the recording was captured.

5   Parties present during the recording.
6   The capture method used (e.g. in a pocket, under a table).
7   The date the recording was purportedly taken.
8   Whether the device is available for analysis.
9   Whether there is any further information available in relation to the device if it is not available.

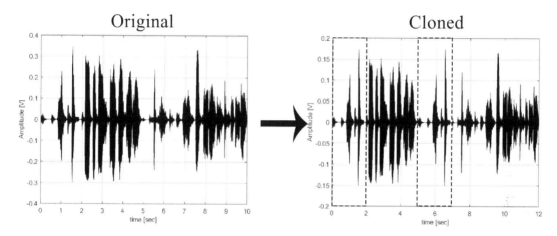

Figure 11.5 Clone function in which the region within the broken lines is cloned and inserted at a different point within an original signal, thus extending the duration by 2 seconds

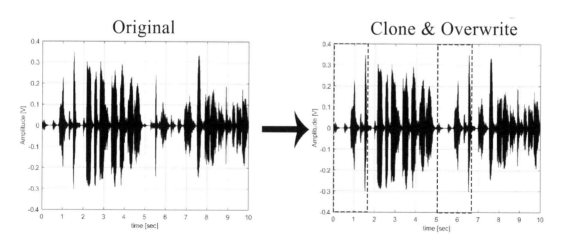

Figure 11.6 Clone function in which the region within the broken lines is cloned and inserted at a different point within an original signal, overwriting the data in this region. The duration, therefore, does not increase.

10 The reason for the authentication (such as believed to be from a different device, believed editing has taken place).

11 The known chain of custody of the provided exhibit, up to, and including the time it was submitted for an examination.

Once this information has been provided, and depending on the responses, it may be that steps can be taken to request the original device and to image the original audio file in a forensically sound manner. It also allows for the user manual pertaining to the purported device to be reviewed, and the purchase of a replica device to create exemplar recordings for comparisons to be made against.

The actual recording device which made the recording should only be used to create exemplar recordings with written permission from the instructing party, as the use of this will alter the evidence device (especially if it does not contain external storage memory) (SWGDE, 2018). If the evidence recording was captured on memory which can be physically removed from the device, this should be removed and safely stored. Newly formatted removable memory can then be inserted and exemplar recordings captured with no risk of altering the original evidence.

Initial assessment

Once the recording has been imaged, and as much information as possible gathered from both the instructing party and through other means (such as reviewing the associated user manual), an initial assessment of the file can be performed and findings documented. For example, is the file a video or standalone audio recording? Is it encoded in a standard format or does it require a proprietary player for playback? If it does require a proprietary player, what are the capabilities of the player? Does it allow recordings to be edited and exported? (See Koenig and Lacey, 2012a.)

All of the above can then be used to determine the direction of the analysis. If the source is known and the device is available, the first task may be to create exemplar recordings. If a proprietary player is required but has not been supplied, the first task would be to locate the player by contacting the instructing party or through independent research into the software required. The results should be documented in working notes, for which Table 11.1 provides a simple example.

Table 11.1 Purported capture device working notes

Purported Capture Device Make
Purported Capture Device Model
Purported Capture Device Operating System
Purported Capture Device Firmware
Purported Capture Device Capabilities (e.g. sample rates, codecs, editing options)
Purported Capture Environment
Purported Capture Contents
Purported Capture Date & Time
Known Chain of Custody

Table 11.2 Overview of methodologies available

Purported device known?	Available material	Methodology
Yes	Purported capture device	Comparison against original and edited exemplar recordings of matching duration, similar capture environment and similar content using purported capture device.
Yes	Purported capture device of the same make and model	Comparison against original and edited exemplar recordings of matching duration, similar capture environment and similar content using purported capture device of same make and model.
Yes	Exemplar recordings from purported capture device of same make and model within database	Comparison against original and edited exemplar recordings of the purported capture device of same make and model
No	Exemplar recordings from a variety of devices and editing software within a database	Comparison against a database of recordings from known devices and editing software to determine if there is a match

Some initial conclusions may also be drawn where applicable, for example, if the format is proprietary and the software has no editing features or options to export recordings, the opportunities for editing are limited. On the other hand, if the recording is in an MP3 format and no additional information is known, it opens up a vast number of possibilities as to the provenance of the recording and any processing which may have occurred. Table 11.2 provides a general overview of the methodologies which should be followed depending on the information and material available.

In cases where no information is known relating to the purported device and no match is found against original or edited exemplar recordings from a database, the result of the examination must be inconclusive, or the case declined (ENFSI, 2005).

Proprietary recordings and video

As recordings of these types are generally in a format which cannot be opened by standard audio tools used for the analysis of the content, there may be no other option but to rewrap or transcode the exhibit. If this is necessary, great care must be taken to preserve the integrity of the audio data to prevent any changes being made. If this is not possible, a caveat must be made within the report and taken into account when performing analysis and drawing conclusions. Any analysis of the file structure and associated metadata must be conducted on the working copy, as once converted, this will change. The best option is to rewrap the audio, which essentially creates a bit-stream copy of the audio stream and wraps it in a new container. Exporting from a proprietary player or transcoding will change the composition of the data, so great care should be taken when performing any analyses to ensure potential artefacts are not due to the conversion process.

For audio provided with a proprietary player, a direct export from the said player may be the only option available. In cases where this option is not available, considerations must be made as to whether the authentication should be aborted or the audio captured from the sound card. If the latter option is selected, it is essential the method has been validated, and the limitations of the examination are understood and documented in the report.

For proprietary video exhibits, the audio stream should be rewrapped in an audio container, thus ensuring the integrity of the data is maintained. This ensures the recording can be opened in workstation software for critical listening and other specific analyses to be performed. Conclusions concerning the file structure, metadata, and format should be made using the working copy of the video exhibit file.

Exemplar recordings

Long before the capture of the first digital audio recording, Albert Einstein stated, 'Everything is relative.' As the measure of the authenticity of a recording is its 'consistency with an original recording,' there needs to be a relative recording for the exhibit to be consistent with. It may be that there are apparent traces relating to editing software within an exhibits' file structure, but without a recording exported from the editing software for comparisons to be made, the conclusions will be limited as an exhaustive examination cannot be performed. The most robust conclusions can only come through objective comparisons of various elements of the exhibit against exemplar recordings.

The way in which exemplar material is used is dependent on the scenario, of which there are essentially two:

1 Cases where a recording is purported to be from a specific recorder.
2 Cases where there is no information available in relation to the purported recording device.

Cases where a recording is purported to be from a specific recorder

In this type of case, there are two options in obtaining exemplar recordings for comparison: either use a device of the same make, model, and operating system/firmware as the original, or use the purported capture device itself (SWGDE, 2018). The former is the ideal scenario as it ensures the recording device remains in its original state, but it also limits some of the analyses, and due to unavailability, this may not be possible. Obtaining exemplar recordings from the instructing party which they accept as original may appear to be a viable option, but it must be considered that they too may be edited. The only exemplar recordings used should be those which can be trusted to be an accurate representation of what they purport to be, and the only real way this can be ensured is by the creation of exemplar recordings by the laboratory. If the provenance of a recording cannot be confirmed, and thus trusted, it can be reasoned that they cannot be considered to be reliable for use as an exemplar.

In terms of capturing exemplar recordings, they should match the exhibit as closely as possible. For format, specifications, and duration, this is not difficult, but for obvious reasons, content-wise it is. Attempts should be made to mimic the acoustic content as far as possible, so if the recording contains speech captured within an indoor environment with a small amount of reverberation, then these should be the conditions replicated. This ensures you are comparing apples against apples instead of against bananas.

Not only should original recordings be created, but also ones which have been edited within the device. This is for two reasons. First, it prevents incorrect conclusions being drawn if a recording edited within the device is similar to the original recording, so leaves no traces of editing. Second, it strengthens the conclusions. For example, if an exhibit is inconsistent with an original recording but consistent with an edited recording from the same device, the conclusion is stronger than simply concluding that it is inconsistent with an original (or vice versa). It may also be necessary to perform various tests to mimic any artefacts or events which are found within the exhibit. For instance, if there is a break in the recording, recordings should be made which contain pauses, either through the manual application of a pause button or through the use of a voice-activated option which stops the recording whenever the magnitude of the audio input drops below a specific threshold. Tests could also include powering down the device during recording, removing the memory card, and any other methods which may have potentially caused the breaks in the recording. It is only through the capture of exemplar recordings and by using all of the available options for comparison that the most exhaustive conclusions can be drawn.

A case where there is no information in relation to the recording device

In terms of the second scenario, this is increasingly common due to the ease with which digital information can be shared. An example would be somebody attempting to bribe an individual by sending a recording via email but refusing to provide information as to the method used to obtain it. Another would be the audio stream of a video uploaded to the internet.

In this case, there may be traces within the recordings metadata which provide clues as to the make and model of the capture device or software. If these are found, exemplar recordings can then be created with the documented device or software as in the first scenario. If no information is available, then analyses will be limited, and this limitation should be stated in the report. The authentication should focus on the purported method of capture (for example, it is claimed to have been captured over a telephone) and the consistency of the recordings with known audio editing software and known editing techniques. In order to make comparisons, exemplar recordings from editing software and various capture methods are essential, which leads neatly into the next section.

Database

An extension from the creation of exemplar recordings is the creation of a database of exemplar recordings so further comparisons can be made against exhibits. This is not only useful when the device is known but unavailable, but also when no information is known about a recording, as it may be that it matches one in the database. The type of recordings should include, but not be limited to:

1   mobile phone recordings (audio, video, original, and edited);
2   portable recording devices (audio, video, original, and edited);
3   recordings exported from audio editing software;

4　recordings exported from video editing software;
5　streaming website downloads (audio and video);
6　social media platforms (audio and video);
7　mobile phone messaging service transfers (audio and video).

Granted that it is an impossible task to obtain exemplar recordings from even a small percentage of the devices available on the market due to the resources it would require, and the rate at which new devices and software are released, having some form of exemplar material in the shape of the most popular formats is not.

Ideally, recordings should be obtained in all possible formats from each source, in conditions which mimic real casework. The compiling of a database will then afford the examiner the most scientific and objective conclusions possible, and the option to take the database compilation further into areas such as research and coding of algorithms to perform comparisons autonomously.

Analysis

Analyses can be classified in different ways, each of which can help in constructing a methodology to approach an authentication.

The first branches of classification are those of container and content analysis. Container analysis pertains to the examination of the container in which the audio data resides, and includes reviewing both the file structure and metadata. Content analysis focuses on the bytes which form the raw audio content (Zakariah et al., 2018).

The second set of branches is related to global and local analysis of the audio stream (Table 11.3). Global analyses are those which provide results pertaining to the audio data in its entirety, such as traces of a specific device. Local analyses are those which provide results pertaining to specific locations, for example, to show the points within the time domain at which editing has occurred. Finding local editing points should always be the primary objective, as they provide the highest level of detail with regards to editing and the strongest support for any global findings. If a local edit point cannot be found within a recording that shows inconsistencies with an original recording through global analyses, it is not an indication that editing has not taken place, just that there are no local traces detected. As the adage goes, 'Absence of evidence is not evidence of absence.'

Critical listening

Critical listening is just that, listening to the audio content in a critical manner. It should be one of the first, if not the first, analysis performed in any authenticity examination. As critical listening is subjective, it is not an analysis which has a considerable weighting on the conclusion in and of itself but is essential in providing direction for the other objective analyses which follow. No inferences should be drawn from this stage as it may create an expectation bias for analyses performed later on. Instead, hypotheses should be prepared that can then be tested, thus ensuring empirical, objective results (Grigoras et al., 2012).

The most significant advantage of critical listening is that the human auditory system has the ability to identify specific sounds buried among noise and extraneous signals which other

Table 11.3 Overview of analyses available

Area	Type	Domain	Analysis
Container	Global	N/A	Filename
			Filename Extension
			File Format
			File Header
			File Footer
			File Structure
			Metadata
Content	Global	Frequency	Spectrogram
			Differentiated Sorted Spectrum
			Compression Level
			LTAS
			LTASS
			MDCT
			Power Spectral Density
		Time	Critical Listening
			Waveform
			Zero Padding
			DC Offset
			Quantisation Level/Bit Depth
			Stereo Phase
	Local	Frequency	Spectrogram
			MDCT
		Time	Critical Listening
			Waveform
			Windowed Power Spectral Density
			Stereo Phase
			ENF
			Windowed DC Offset
			Windowed Bit Depth
			Butt Splice

analyses are not designed for, or are not capable of detecting (Everest, 2006). The disadvantage is its subjective nature, as what to one person may be perceived as an edit point may be heard as an acoustic impulse by another. Although reducing the subjectivity of critical listening is difficult, the reliability of any opinions drawn can be improved by having an understanding of artefacts related to digital audio encoding and their potential causes. For example, to the untrained ear, a change in the broadband noise in a recording of a phone call may be considered to be caused by an edit, but to an examiner with an understanding of techniques used by encoding algorithms pertaining to phones, the transition may be explainable as an artefact related to voice activity detection (VAD). Another disadvantage is that the ear cannot detect a vast number of tampering artefacts, and so other methods are required in order for them to be exposed.

The critical listening examination should be performed using a bit-stream copy of the original recording wearing closed-back, over-the-ear headphones to prevent convolution of the audio with the environment. If either of these prerequisites were to be ignored, there is a potential for the clarity to be reduced, and artefacts introduced which are not contained within the exhibit recording. It may also mask others which are.

The exercise should first be performed without any visual aid such as a spectrum to prevent biases or distractions from the audible events. It can then be repeated while viewing a spectrogram and waveform, and further observations relating to any correlation between the acoustic and visual events documented. It is often the case that the process must be repeated several times as there are a variety of areas which must be considered, and it is likely to be very stop-start rather than a straight run through. When potential artefacts are detected, it will necessitate the section being replayed to listen even more critically, confirm any findings, and to note the offset at which they occur so further analysis into such can be performed.

The role of critical listening is essentially a pre-analysis, and findings can be divided into four regions, namely, preliminary overview, record events, background sounds, and foreground information (Koenig, 1990). The overview is always performed first, and the order of the others informed by findings from the first stage. Each area will now be explained separately, and further details of the potential artefacts within each provided.

Preliminary overview

This is the first critical listening exercise, in which the recording is reviewed in its entirety to document the chronology of events, including stop-start occurrences, background noise differences, analogue-like record events, playback speed transitions, signal dropouts, and other variations which require further study.

SIGNAL LOSS

Instances of the audio abruptly cutting off should be noted, and the possible reasons for this investigated. Many devices do have an option for the user to pause the recording, and for some devices, this can be controlled using a threshold, stopping once the signal drops below a certain amplitude. When the magnitude increases and exceeds the threshold, the recording resumes. There are various considerations which can be made to aid in determining the cause for signal dropouts. For example, when a recording is paused using a button on the device, there may be an audible click before the pause. For a recording which stopped due to an automatic function, the recording would likely decrease in magnitude to the same extent before each pause. All of these areas would require further analysis of the purported capture device and the creation of exemplar recordings.

INSTANTANEOUS CHANGES TO THE BROADBAND NOISE

Broadband noise is that which extends across the frequency spectrum and is often low in amplitude. It is caused by the general background ambience of the capture environment, quantisation noise, and the internal electronic components of the recording device. If a section from another recording made using a different device in a different environment is spliced into a recording, or even a section from a different region of the same recording is cloned, there will likely be a subtle change to the broadband noise. There are also obvious innocent explanations, such as the movement of the device or speaker, changes to the environment, changes to the gain of the device, and the introduction of an electrical, magnetic field.

INSTANTANEOUS CHANGES TO THE SPEAKER TO NOISE RATIO

For recordings in which the subject to microphone geometry is constant, and there are no changes to the environment, it would be expected that the speaker to noise ratio would be consistent throughout. If there were a change to either of these two factors, the transition would be relatively slow, such as a voice which gradually increases in magnitude, due to the time it takes for the subject or microphone to move. Obviously, there are exceptions, but in general, sudden and abrupt changes to the speaker to noise ratio would generally be inconsistent with this.

Record events

Record events are those which can be considered to be caused by a specific event associated with the recording, such as a pause, presence of a dial tone, or telephone call cut off.

PAUSE EVENTS

Pause events refer to the operation of stopping and restarting recording by a user during capture. They may be perceptually evident, for example, undercover officers commonly provide context to a recording by stating that 'the recording will now be paused' or 'stopping recording to preserve battery,' or for an officer performing an interview to state that they are 'now pausing the interview' for one reason or another. Others may not be so obvious, such as when there is an abrupt signal loss for a short period or an instantaneous transition in the recording content. The traces left by a pause event vary in amplitude, shape, and transients, not only between the recording device make-model and the format but also between seemingly identical units. For this reason, all events should be compared against exemplar test recordings made using the purported capture device when possible. As different types of pause events have been shown to produce different results, these exemplar recordings should consist of both short pause (pressing pause followed by resuming recording instantly), and long pause events (pressing pause, waiting a period of time, and then resuming the recording). It is essential for all possibilities to be exhausted, as similar artefacts can be caused by numerous functions, including stop, start, sound activation detection, splicing, deletion, and file concatenation.

CONTENT CONSISTENCY

Depending on the content, there are certain events which would be expected to be audible. A good example is the capture of a telephone call. As telephone calls are anticipated to contain dial tones preceding the call, and some form of cut-off once the call is ended, the presence or absence of these clues should be noted. Further information in relation to the system used would then be required to determine the capabilities of the system and find answers to questions such as 'Does the recording begin once the number is dialled, once somebody answers, or is it manually activated by the operator?'

Background sounds

This section relates to all background sounds, including unnatural changes, repetitive noise sources, music, recording device noise, hum, mic handling, and reverberation.

INSTANTANEOUS CHANGES TO BACKGROUND NOISE

Similar to the other instant changes, it is unnatural for background noise within a recording to change with no gradation or audible movement, generally speaking. There are obvious exceptions, such as the switching on of a television or background music, but these should be obvious to the examiner. If the background noise instantly changes from what appears to be an outdoor environment to a loud indoor venue with music playing and numerous voices, then it is plausible this is caused by an edit point. HVAC systems are also a common source of background noise, which may emit a stationary tone. This is often clearly visible within a spectrogram, so any changes in relation to such should be subject to further analysis.

CLICKS AND POPS

There are several causes of clicks and pops, the source of which must be considered for each independently. Cause for concern with regards to editing stems from the fact that when two audio samples which were not recorded from a natural acoustic event meet, there will be an unnatural amplitude transition between them, resulting in an audible click (unless interpolation or cross-fading is applied). In order to determine the cause of the clicks and pops, the following should be taken into consideration. First, does the artefact contain a reverberation tail? If this is the case, then it can be surmised that it must be an acoustic event, as there would be no form of decay envelope if it were an edit point. Second, does the artefact occur across a single word? For example, 'coke' … 'click'… 'aine.' For a would-be manipulator to decide the centre of a word is a good place to perform an edit would be possible, but highly implausible. Therefore, it can be surmised that this is likely an artefact caused during encoding. With that being said, it must be kept in mind that audio editing does not have to occur in a single linear dimension. Spliced sections can be laid over existing audio, something that should always be taken into account when formulating hypotheses and drawing conclusions.

DISTORTIONS

Distortions are sounds which are a product of the audio encoding process rather than the acoustic environment. The most common are those caused by clipping, which results in audible distortion within the clipped regions. Although clipping may not be a direct effect of an edit, if there are audible distortions consistent with clipped samples, but no visibly clipped regions within the spectrogram, the possibility exists that post-processing of some form may have occurred since the original capture. The innocent application of normalisation could cause this in a crude attempt to remove the distortion, or a malicious splice from a region within a clipped recording into a non-clipped one where some form of gain reduction has been applied.

REVERBERATION TAIL CUT-OFFS

As sounds contain a direct signal followed by highly correlated reflections, an edit point which occurs during the reflections can cut the decaying tail short, resulting in a distinctive artefact. The more reverberation present, the more apparent the reverberation tail cut-off will be. For example, it would be extremely noticeable if this were to occur on the reverberation tail of a recording in a cave but would be much more subtle on a recording within a domestic environment. This further demonstrates the requirement for critical listening using suitable equipment in an appropriate laboratory space.

INSTANTANEOUS CHANGES TO THE PROFILE OF THE REVERBERATION

The reverberation profile can be defined as the character of the indirect reflections and the amount of attenuation, which pertain to every acoustic event captured (except those captured within an acoustically dead space such as an anechoic chamber). Although not always measurable, the human auditory system is highly sensitive to reverberation and changes to them, but as every sound we hear contains reverberation to some degree, we may not fully appreciate it without making a conscious effort to do so. If possible, the reader is encouraged to spend a few minutes in an anechoic chamber to understand the impact of reflections on sound and how strange a lack of reverberation can be.

Recordings captured in a single environment in which there is no movement of the microphone would be expected to have a consistent reverberation profile across the length of the recording. Now, imagine a recording in which somebody is talking, and the pre-delay is relatively large (for example, in a church-like environment), and within milliseconds is extremely short, without any apparent cues of movement. Although technically possible, it is a highly unnatural occurrence. There are occasions for which it is explainable, for instance, when the microphone is moved closer to the sound source, the magnitude of the direct sound may mask the indirect reflections, but even then there would be expected to be audible reverberation tails in the spaces between sentences. If the capture device is covered in some way, specific frequencies may be attenuated before entering the microphone, but the reverberation profile would still be expected to be audible to some extent. Similar to speaker to noise ratio changes, any changes to the reverberation profile would be expected to be a natural transition over a period of time rather than instantaneously.

Foreground information

This consists of any sounds which are considered to be in the foreground, including voices, high magnitude sounds, and contextual information.

INSTANTANEOUS CHANGES TO SPEAKER LEVELS

There are essentially two elements to consider: magnitude changes of speech pertaining to a single voice and magnitude changes between two or more voices. Extreme caution must be taken when analysing the magnitude of voices as this is probably the area with the highest number of innocent explanations due to the dynamic range of the human voice. In fact, it is *expected* that in the average conversation, the magnitude of voices will rise and fall to some degree. Without this range, the result would be an extremely monotone (and boring) dialect. Dynamic range adds to the character and distinctiveness of a voice, and even though there are constant changes to the voice, these are all within a natural range. For example, if quiet is considered to be 1 on an arbitrary scale and 10 is screaming at the top of the lungs, a voice may move between levels 4 and 6 for the duration of a conversation. Speech that suddenly changes from a level 2 to a level 10 for a sentence and back to a level 2 would not generally be a natural occurrence, but may be a sign that a sentence from another recording was spliced into the recording with no attempt on behalf of the manipulator to match the amplitudes of the voices. Obviously, if somebody decided to shout a sentence, it would be odd but plausible, so considerations need to be made on a case-by-case basis, taking into account all other factors and cues provided, including the perceived level of vocal effort being applied by the speaker.

INSTANTANEOUS CHANGES TO THE VOICE PROFILE

The profile of a voice is characterised by a myriad of factors, such as the average fundamental frequency, the timbre, dynamic range, and other areas covered in earlier chapters. Models are built on a multitude of these elements for speaker recognition algorithms, owing to the uniqueness of the human voice. When a number of these factors change, this can be audibly detected, and with voice synthesising technology becoming a reality, any changes should be noted in preparation for further analysis. The transmission channel of the recording device can also cause changes to the profile of the voice due to factors such as sampling rate and the frequency response of the microphone. The potential, therefore, exists that a change is due to a section being spliced into a recording of the same voice, but captured using a different recording device. As with all critical listening analyses, there may also be innocent causes such as the microphone to speaker geometry, which is further evidence as to why critical listening should not be used to provide conclusions in and of itself, but instead to find areas for further analysis and cross-correlate with findings from these analyses.

CONVERSATION CHRONOLOGY

The general flow of a conversation is to move from question to answer, and from topic to topic in a logical manner in which the previous sentence or paragraph relates to the next. If the chronology appears to be somewhat odd, for example, a completely unrelated topic bookmarked by a single topic, then this should be noted, and further examination of the area performed.

Critical listening documentation

The simplest way to ensure all points have been considered is to use a checklist. Table 11.4 is an example of this, in which the results can then be assimilated into a report.

Table 11.4 Critical listening checklist

Preliminary overview	
Feature	*Notes*
Clicks and pops	
Signal loss	
Distortion	
Instantaneous changes to broadband noise	
Record events	
Pause event	
Dial tone / Call cut off	
Background sounds	
Reverberation tail cut-offs	
Instantaneous changes to reverberation profile	
Changes to background sounds	
Instantaneous changes to background noise ratio	
Foreground information	
Instantaneous changes to speaker levels	
Instantaneous changes to speaker to noise ratio	

Considerations for video

Although the aforementioned considerations which form the critical listening stage of an examination apply to both standalone audio recordings and video recordings, there are additional aspects which pertain only to video.

SYNCHRONISATION

The degree of synchronisation between lip movement captured within the video stream and speech within the audio stream is one of the more obvious considerations during a critical listening exercise of video. This may not be possible if the content does not feature speech, or the lips of the subject talking are not visible within the video capture.

All events within the imagery which create a sound wave should also be reviewed. For example, a gunshot would likely show a muzzle flash which coincides with an audible pop. In drawing conclusions with regards to non-synchronised events, a quantitative measurement should be obtained where possible to reduce subjectivity. This can easily be performed by calculating the offset of transient sounds and corresponding visible events. For instance, the time elapsed between doors seen to be closing and the sound associated with such should be documented at various points throughout the recording to determine if there is any unexpected drift which may be consistent with editing or the attempted synchronisation of multimedia streams from two different sources. Imagine a recording that is fully synchronised when reviewing events towards the beginning and end of the recording, but there is an offset of 2 seconds between the visual events and sounds between these points. This is difficult to explain through drift, because if this were the case, poor synchronisation would also be expected towards the end of the recording, so further analysis of the affected area should be performed. Another example would be a recording which begins fully synchronised but drifts apart further and further throughout the recording. If an audio stream was spliced into a video recording, timing differences between the sample rate and frame rate might be one potential explanation for the drift, but it could also be due to internal timing issues with the capture device. This evidences the requirement to always create hypotheses and cross-correlate findings of various analyses, as many artefacts can have a number of causes which are both innocent and nefarious in nature.

CONTENT MISMATCH

This pertains to a perceptual mismatch between the audio and video streams, for which there are a number of areas to be considered, such as:

- *Background noise.* Is the background noise consistent with on-screen events? For example, visual events showing the inside of a property, but background noise from a car engine.
- *Reverberation.* Does the audio appear to have a different reverberation profile than would be expected for the environment captured on screen? For instance, a recording captured in a field would not be expected to have a reverberation profile similar to a church or large room.
- *Voice levels.* Are the voice levels consistent with the visual subject to microphone geometry? For example, is the voice of the subject furthest from the microphone of a higher magnitude than the voice of the subject closest to the microphone, when both appear to be using the same amount of vocal effort?

Critical listening breaks

Critical listening should be performed at a playback volume which will not rapidly introduce ear fatigue, a condition which reduces the sensitivity of the human auditory system and possibly results in misinterpretation or certain events going unnoticed.

As the auditory system becomes tired, it is common for listeners to increase the volume of playback to achieve the same perceived level as when they first began listening. This is an obvious sign of fatigue and one that indicates a break is required, as not only will the analysis be sub-optimal, but the repeated practice of increasing the level over extended periods will result in permanent and irreversible hearing damage.

Regular listening breaks should, therefore, be taken to aid in preventing ear fatigue. A good rule of thumb is a 5-minute break every 30 minutes to allow the auditory system to return to its natural level of sensitivity.

Filename

Theory

The vast majority of recording devices will have a standard naming format, but many also give the user the option to enter a filename for a recording. This can be once the recording has been captured (and before saving the file) or through the user navigating to the specific option to change a filename within the capture device software.

Application

One of the simplest, but least reliable forms of analysis, is to compare the exhibit filename and its relationship with the naming scheme used by the purported capture device (SWGDE, 2018). A recording named 'editedrecording.wav' would most likely not be the original recording, as it would be tough to imagine a recording device using a naming scheme of this kind.

If the purported capture device or an identical make and model running the same software version is available, comparisons can be made between the filename format of the exhibit against exemplar recordings. Although this will provide nothing conclusive as (1) people have a habit of changing filenames regardless of whether the recording has been edited or not, and (2) the ease with which a filename can be changed means the savvy manipulator would ensure the name has been changed to reflect the file naming format of the device from which the recording is purported to have come from, it should still be documented in working notes as it may cross-correlate with information found later in the examination. It would primarily be supporting evidence which does not influence the conclusion. It can also be of use when no information pertaining to the purported capture device is available, as comparisons can be made against exemplars within a multimedia database.

Filename extension

Theory

A second basic analysis is to view the filename extension of the recording, as these too will be standardised by the capture device. Filename extensions are used by operating systems to ensure files are opened in suitable software to ensure the data within is correctly interpreted and accessible to the user. For example, software designed for the playback of audio would not be able to open a file containing only text, and thus text documents have a.txt filename extension, ensuring they are opened in software which can read this type of data and present it to the user. Similarly, specific proprietary audio formats will have filename extensions which are not recognised by the operating system, and it therefore may not be possible to open the file without a specific proprietary player.

Application

It may be that the recording has an MP3 file extension, but is purported to have been captured using a device which only stores audio in an AAC format recordings. Again, this should in no way be considered conclusive, and great care must be taken in the interpretation of this information. With that being said, it should be documented as a further investigation of the file structure can confirm or dismiss any concerns over the extension. For example, if the file is found to have an M4A structure, it may be that the filename extension was innocently changed in an attempt to achieve playback, not understanding its role. Many different formats also use the same filename extensions, for example, a WAV extension can indicate a WAV PCM encoded audio recording, but the extension is also shared with audio data encoded using ADPCM, DVI/IMA, a-law and mu-law formats. For this reason, a good understanding of the various audio codecs, their associated file containers and their filename extensions is essential, as well as the capture and storage of exemplar material.

File location

Theory

In a similar vein to the previous two analyses, devices also store files in specific default locations within their file structures.

Application

Although this analysis is not possible if the device has not been provided, when made available, it does allow considerations as to the consistency of the location in which the exhibit is located against that of the default storage location for recordings captured with the purported capture device (National Institute of Justice, Office of Justice Programs, 2004). Other areas to consider are whether the area in which the recordings are stored can be accessed through other means, such as direct cable connectivity with a computer or over a network. If it is found that the area is not accessible to a user, it limits the possibilities in relation to potential manipulation of the exhibit, as it must have been captured by the device on which it is stored, and thus can only have been edited within the device.

File structure

Theory

As discussed in Chapter 6 on audio encoding, if the recording is in a standard format, there is a requirement for the files to be structured within a specific way based on audio coding standards to ensure a recording can be accurately decoded (Grigoras and Smith, 2017; Koenig and Lacey, 2012b). For example, the presence of a 'RIFF' ASCII offset within the header structure of a file (Figure 11.7) is a requirement of WAV PCM files (Microsoft Corporation and IBM Corporation, 1991). Similarly, a 'QT' ACSII offset (Figure 11.8) is a requirement of files which use the Apple QuickTime container format (Apple, 2016). A hex viewer/editor is required to view the file structure, of which there are multiple options, both free and commercial.

Application

The predictable nature of file structures allows robust comparisons to be made between those of exhibits and those of original and edited recordings from a purported capture device. Proprietary formats require a more critical analysis of the file structure as there is no published standard to refer to. The creation and analysis of exemplar recordings from the purported device are therefore essential to aid in understanding potential differences in file structure between original

```
Offset(h)  00 01 02 03 04 05 06 07 08 09 0A 0B 0C 0D 0E 0F  Decoded text
00000000   52 49 46 46 24 5D A9 44 57 41 56 45 66 6D 74 20  RIFF$]©DWAVEfmt
00000010   10 00 00 00 01 00 02 00 44 AC 00 00 10 B1 02 00  ........D¬...±..
00000020   04 00 10 00 64 61 74 61 00 5D A9 44 00 00 00 00  ....data.]©D....
00000030   FF FF 00 00 FE FF 01 00 FD FF 02 00 FC FF 02 00  ÿÿ..þÿ..ýÿ..üÿ..
00000040   FC FF 03 00 FB FF 03 00 FA FF 04 00 F9 FF 05 00  üÿ..ûÿ..úÿ..ùÿ..
00000050   F8 FF 05 00 F8 FF 06 00 F7 FF 06 00 F6 FF 07 00  øÿ..øÿ..÷ÿ..öÿ..
00000060   F5 FF 08 00 F4 FF 08 00 F4 FF 09 00 F3 FF 09 00  õÿ..ôÿ..ôÿ..óÿ..
00000070   F2 FF 0A 00 F1 FF 0A 00 F0 FF 0B 00 EF FF 0B 00  òÿ..ñÿ..ðÿ..ïÿ..
00000080   EE FF 0C 00 EE FF 0C 00 ED FF 0D 00 EC FF 0D 00  îÿ..îÿ..íÿ..ìÿ..
00000090   EB FF 0E 00 EA FF 0F 00 E9 FF 0F 00 E8 FF 10 00  ëÿ..êÿ..éÿ..èÿ..
```

Figure 11.7 Hexadecimal/ACSII representation of a WAVE audio file containing the expected RIFF offset
Source: Captured from HxD.

```
Offset(h)  00 01 02 03 04 05 06 07 08 09 0A 0B 0C 0D 0E 0F  Decoded text
0021E550   21 6F 70 00 00 01 53 75 64 74 61 00 00 00 20 64  !op...Sudta... d
0021E560   61 74 65 32 30 31 39 2D 30 31 2D 31 30 54 31 36  ate2019-01-10T16
0021E570   3A 30 33 3A 31 32 2D 30 37 30 30 00 00 01 2B 6D  :03:12-0700...+m
0021E580   65 74 61 00 00 00 00 00 00 00 22 68 64 6C 72 00  eta......."hdlr.
0021E590   00 00 00 00 00 00 00 6D 64 69 72 00 00 00 00 00  .......mdir.....
0021E5A0   00 00 00 00 00 00 00 00 00 00 FD 69 6C 73        ...........ýils
0021E5B0   74 00 00 00 BC 2D 2D 2D 2D 00 00 00 1C 6D 65 61  t...¼----....mea
0021E5C0   6E 00 00 00 00 63 6F 6D 2E 61 70 70 6C 65 2E 69  n....com.apple.i
0021E5D0   54 75 6E 65 73 00 00 00 14 6E 61 6D 65 00 00 00  Tunes....name...
0021E5E0   00 69 54 75 6E 53 4D 50 42 00 00 00 84 64 61 74  .iTunSMPB...„dat
0021E5F0   61 00 00 00 01 00 00 00 00 20 30 30 30 30 30 30  a........ 000000
0021E600   30 30 20 30 30 30 30 30 30 38 34 30 20 30 30 30 30  00 00000840 0000
0021E610   30 32 35 32 20 30 30 30 30 30 30 30 30 30 30 33  0252 00000000003
0021E620   35 36 44 36 45 20 30 30 30 30 30 30 30 30 20 30  56D6E 00000000 0
0021E630   30 30 30 30 30 30 30 20 30 30 30 30 30 30 30 30  0000000 00000000
0021E640   20 30 30 30 30 30 30 30 20 30 30 30 30 30 30 30  00000000 000000
0021E650   30 30 20 30 30 30 30 30 30 30 30 20 30 30 30 30  00 00000000 0000
0021E660   30 30 30 30 20 30 30 30 30 30 30 30 30 00 00 00  0000 00000000...
0021E670   39 A9 74 6F 6F 00 00 00 31 64 61 74 61 00 00 00  9©too...ldata...
0021E680   01 00 00 00 00 63 6F 6D 2E 61 70 70 6C 65 2E 56  .....com.apple.V
0021E690   6F 69 63 65 4D 65 6D 6F 73 20 28 69 4F 53 20 31  oiceMemos (iOS 1
0021E6A0   32 2E 31 2E 32 29                                2.1.2)
```

Figure 11.8 Hexadecimal/ACSII representation of an Apple QuickTime audio file containing the expected 'UDTA' offset
Source: Captured from HxD.

and edited recordings. When you consider the amount of work required to create a standard, in terms of both the knowledge and time, most companies will not reinvent the wheel as it would be too immense a task. Many proprietary formats will, therefore, use a standard codec to encode the audio data and store it within an encrypted format. In general, it is the devices designed for security purposes such as dial-in probes and undercover recordings that will not conform to a common standard. Analysis of these types of recordings can be performed through manual and automated searches of the data to identify the repetition of specific bytes which can indicate the block structure. For example, a region which is repeated every 512 bytes is likely to indicate the start of each frame within the payload section of a file. Any deviation found from this structure may indicate corruption or potential manipulation of the data.

The audio data rate can also be verified through the following equation:

$$\text{Bit rate (bits per second)} = \text{total bits of audio content/duration of recording in seconds}$$

The bit rate can then be converted to bytes if required and compared against the capabilities of the purported capture device. For instance, if the alleged capture device uses a rate of 64 kb/s, but the audio bit rate was calculated to be 256 kb/s, this would be inconsistent with an original recording from that device.

There are also regions within file structures for developers to add textual information, such as in the ID3 tag that is generally associated with the MP3 format. This can include:

- start time, end time, and date of the recording (taken from internal date-time settings of the device);
- start time, end time, and date of each region (for example ,if paused. again this is taken from internal date-time settings of the device);
- channel configuration;
- sample rate;
- duration;
- quantisation scheme;
- compression rate;
- capture device make;
- capture device model;
- capture device software version;
- capture device firmware version;
- capture device serial number;
- protection data such as CRCC or checksums.

Where the location of the above is not apparent, an exemplar recording can be created with known variables, and this then used to determine the location of a specific byte. For example, the date and time settings of the exemplar device can be documented before an exemplar recording is created and searches of the file structure performed, which pertain to the previously recorded date and time, in both ASCI and hexadecimal formats. If the year is set as 2000, searches could be performed on the ASCII data for this number, and searches within the hexadecimal representation for '7D0' (2000 in hex). If the date is found, the corresponding location can then be analysed in the evidence recording to find data relating to the capture time using markers from the surrounding bytes. All metadata contained within the file structure can then be used to correlate and verify with other known data, for example, the number of channels or the sample rate of the recording.

All of the above can result in file structures that are highly predictable as they are first of all specific to encoders and containers, and, second, they are specific to recording devices. Another layer of predictability lies in the fact that devices use specific encoders. This can also allow conclusions to be rendered without the requirement for an exemplar recording, for example, an ID3 tag within a file which has a.wav extension would be suspicious as they are most commonly used within MP3 recordings. Other formats are not as simple to analyse. Devices used by law enforcement will often use measures to prevent the manipulation of data at any point in the recording chain, through various methods including:

- limited functionality of hardware (usually only physical buttons for record and stop functions with no option to output audio through a line-out);
- proprietary, and often encrypted formats which can only be accessed using proprietary software provided by the manufacturer of the device;
- limited functionality of the provided proprietary software (no options to edit, replace or delete data, and no options to open recordings in any format other than that used by the device);
- block-by-block verification through a cyclic redundancy check (CRC) or another method to calculate fixed-length checksums for each block, based on the data within the said block. This is then used to verify the block has not changed since capture upon decoding, and any changes to the data will result in a corrupted and likely unplayable recording.

In scenarios where the factors mentioned above are encountered, the examination will likely focus on the capabilities of the hardware and software to determine if editing is actually possible.

A final note should be made in relation to file structure analysis. First, hex viewers are efficient tools for viewing the file structure, but also for editing the file structure. On the plus side, any changes made to the file structure by somebody with no understanding of this may cause the recording to become unplayable. If eight bytes were deleted, the location of all data beyond this would move position by eight bytes and potentially cause problems upon playback as the decoder would not find certain offsets where they would be expected to reside. It is also incredibly challenging to edit audio content with any degree of accuracy while viewing it in a hex format, as it generally does not resemble audio samples as we know them (although there are some exceptions). Changing an ASCII date is much easier for somebody who understands hex structures as it would just involve swapping out the pertinent bytes for new ones, ensuring no bytes are added or removed which could alter the offsets of other blocks. With that being said, date-time information is sometimes stored in other locations of the file which may not be readable in ASCII format, so changes would be picked up as there would be a disparity between the information extracted from different locations within the file.

It must also be considered that the file structure of a recording which has been edited using tools within the purported capture device may contain the file structure of an original recording from the device (Smith et al., 2017). This does not render the analysis useless (as determining whether the recording came from the purported device is the main reason analysis is performed), but caution should be exercised before drawing any conclusions. For this reason, a review of a device's capabilities and features is essential when performing analyses, as it may well be that the file was captured using the device, but understanding that features exist for sections to be deleted or overwritten is essential in arriving at reliable conclusions. This is also the reason exemplar recordings should be created of both original recordings and those edited using the various options available within the software.

File metadata

Theory

Metadata pertains to data about data. In the case of audio forensics, the 'data' is the exhibit recording, and the data about exhibits can be extracted using various software applications, both open-source and commercial. In breaking down the types of data which can be extracted, those which can be found within the file header or footer as ASCII-readable data will not be covered. While the file structure may contain some information, much of it is also stored within other areas of the file which cannot be read without identifying the location of the information and then converting it into a format we can comprehend. Luckily, there are various software suites capable of extracting this information and converting it to a human-readable format. The specifications of the recording will be covered in another section so are also not considered here.

Application

Not all metadata is of value, and in fact, some can be misleading, so each must be understood, thus allowing inconsistencies with an original recording to be exposed.

DEVICE INFORMATION

Companies often use the container to include textual information about the make and model of the software or device with which the recording was encoded, the specific software version used, and the serial number of the device. Some file specifications have areas within the file in which this should be added, for example, the Apple Quicktime M4A 'udta' (user data) atom (Apple, 2016). Some software companies go a step further and include the complete processing history of the recording before it was exported from their software. They include the name and path of the session file, the name (or names if one recording was spliced into another) of the files used with the session (often referred to as 'source files'), and the number of times the session was saved, among other things. The manipulator is likely unaware of the storage of this information and this can provide incredibly damning evidence of editing. The potential even exists for the name of the individual who created the recording to be documented if they have a profile on the computer that was used for the processing. Not only does this type of information provide objective analysis to correlate with other findings, but may also allow the filenames of the source recordings used to create the editing recording to be provided to the instructing party. All of this information may be in the file structure somewhere (dependent on the software used to encode the recording), but using specialist software to extract it can make life much easier.

TEMPORAL INFORMATION

The area with the most significant potential to mislead is that of MAC times. MAC stands for Modified, Accessed, Creation, and contains temporal information, which to the layman may appear to be of significance, and it is, but only if the information is extracted directly from the device in a bit-stream copy. This is because MAC times provide information as to the time and date the recording was first introduced to a digital system (Rappaport, 2012). For example, imagine a recording is captured at 07:00 on the 1st January and questions are raised over its authenticity. A forensic examination is then instructed, and the examiner is given a copy of the recording, but no access to the purported device. They image the recording onto their workstation at 12:00 on the 5th June. This is the C (Creation Date), as it is the first time a file appeared on the system. The recording is then opened in a Digital Audio Workstation (or DAW) at 12:05 on the same day. This is the A (Access Date). When they close the DAW at 12:10 they select save, overwriting the original file (which is extremely poor practice, but performed for the purposes of this example). This is the M (Modification Date). The recording is then put on an external media device and passed to another examiner, who images it to their workstation at 13:00. The previous MAC data from the first workstation is now replaced with that from the new workstation. In this case, it should be clear that this information must, therefore, be treated with caution as it often provides no information pertaining to the history of the exhibit before it entered the laboratory. In a case where the recording is on a device, and the device is seized, it may be of great importance, so understanding how MAC times work is essential to ensuring they are ignored where necessary and taken advantage of when the situation allows.

An important point to note about any temporal data obtained from recordings is the reliability. Date-time settings of devices are more often than not adjustable by the user, and as the date-time information contained within a recording is taken from the date-time used by the device, care must be taken when conclusions are drawn from this data. The most accurate form of discrete temporal information we have is Coordinated Universal Time (UTC), which can be used interchangeably with Greenwich Mean Time (GMT). Although they share the same time, UTC is a time standard, whereas GMT is a time zone, and neither adjusts for Daylight Saving Time (DST). It is for this reason that in the summer months, the UK follows British Summer Time (BST) (Axon, 2019). Without knowing the difference between the date and time on the device and UTC, any analysis in relation to such is limited. Even if the date and time have not been maliciously adjusted, clock drift alone can cause a disparity between UTC and the time on the device. The caveats regarding the reliability of temporal information should, therefore, be made clear when reporting any opinions based on them. One instance in which timing is more reliable is if it is associated with GPS data (see further below), as this is not taken from within the phone, but from a satellite. It is always reported in a UTC format, so considerations must be made for the difference in local time and UTC when drawing conclusions, taking into account any daylight savings with regards to the GPS location data reported. Care must always be taken as this may indicate the time the capture began, the time the capture ended, or the time of the last edit. Exemplar recordings should, therefore, be created and analysis of such performed.

WRITE PERMISSIONS

A field which can generally be ignored when the exhibit is provided as a bit-stream copy is those of write permissions. These are related to the local workstation or system on which the file exists (most likely the one on which the analysis is being performed) and so have no bearing on the case. They should report that the exhibit is read-only if best practices have been followed and all write permissions removed.

GPS INFORMATION

Some devices allow the file to contain embedded GPS data within the metadata, but it is often dependent on whether the user activates this function. Although more reliable than temporal information, care must be taken to determine if the GPS data pertains to the original recording or a modified version, if the recording has been trimmed or copied. As always, the creation and analysis of exemplar recordings should allow for this to be confirmed. Considerations must be made as to the accuracy of GPS location data as it is dependent on both the hardware and the GPS receiver. Still, most current civilian GPS devices commonly achieve accuracy of under 15 metres (Strawn, 2009).

Other data pertaining to the device with which the recording was purportedly captured can include (but are not limited to) the device make, model, model version, firmware/software version, and serial number.

File specifications

Theory

Audio file specifications differ from device to device and software to software. Those reported include the encoder used, the bit rate, the sampling frequency, the bit depth, and the number of channels. With most devices and software, a user will have some control over the specifications, for example, there may be an option within the settings menu which allows the format to be selected or the sample rate to be increased, but these will be dictated by the device.

Application

As the available specifications of a digital audio recording captured from a specific device are limited, comparisons can be made between the exhibit format and those which the device is capable of encoding. For instance, a recording which is captured as a WAV format at 8 kHz could not be an original recording if the purported capture device only allows MP3 capture at 44.1 kHz. Great care should be taken to ensure that an exhaustive study of all firmware options available for the device in question has been carried out, as these can provide unexpected changes to a device's functionality. It may be that the first firmware version of the previously mentioned recording could not store audio in a WAV PCM, but an update added that functionality. In an ideal scenario, only the exact make and model of the purported device should be tested, as even one which is similar but a newer iteration, or from a different region could have completely different functionality depending on the manufacturer's discretion. If this is not possible, extensive research must be undertaken during the analysis stage and any limitations and caveats documented in the report.

In order to make these comparisons, the specifications must first be extracted. Various software is available to perform this task, and just as for hex viewers, they range from free to commercial versions. It is also possible to extract the data manually from the hex viewer in some cases, but this requires both time and an understanding of the hex structure of the specific file format in question.

Amplitude statistics

Theory

Although a number of other amplitude analyses feature statistics, it is those which can be simply calculated from the audio content and presented numerically to which this section refers. The two most prominent are the number of clipped samples (measured in samples) and the peak sample level (measured in dBFS), both of which are covered in previous chapters so the theory relating to these will not be repeated here.

Application

Clipped samples should be reviewed, as recordings which are found to be distorted through critical listening but show zero clipped samples, will require further analysis to determine the source of the distortion. The first stage of this analysis should consist of reviewing the waveform, as a recording which is found to have a peak sample level of an integer value (for example, -1.00 dBFS), as a natural occurrence would be relatively rare. Not impossible, but rare. If this is further combined with audible distortion consistent with clipping, but zero clipped samples are present, then it can be surmised that the original recording has been normalised to -1.00 dBFS. Pre-sets for normalisation (and gain) are generally provided in integer increments, and there is little reason to adjust this value when performing post-processing, so integer peak sample levels are usually (but not always) the result of normalisation. This is a prime example of why cross-correlation of various findings must be used to formulate conclusions. In isolation, not many findings would be very compelling, but the sum of their parts is much more powerful than their individual components.

Other statistics can also be calculated, such as the root mean square (RMS) (average, minimum, and maximum), and the peak-to-peak amplitude value, but these are of little use due to the unpredictable nature of audio recordings. A low minimum RMS is not indicative of anything in and of itself but may have value in some examinations, such as where a number of recordings are provided which are purported to be sequentially captured using the same device, in the same environment, with the same subject-to-microphone geometry. The analysis may then be made with regards to the consistency of the RMS values, for instance, if one recording within the sequence has significantly higher amplitude measurements in comparison to the other exhibits, it may be indicative of a recording captured in different conditions or with a different device. With that being said, any findings with regards to such would still be of little evidential value in isolation and would be used solely in a supportive capacity.

ENF criterion

Theory

ENF stands for Electric Network Frequency and is based on the rate at which turbines rotate to generate Alternating Current (AC) energy for a power grid. The speed of this rotation is approximately 50 times per second (50 Hz) in Europe, Asia (except Saudi Arabi), Africa (except Liberia), Australia, and parts of South America. In North America and other parts of South America, this rate is 60 times per second (60 Hz). Japan is an exception as it uses both (Nicolalde Rodriguez et al., 2010). In reality, these rates are in a constant state of fluctuation due to the difference between the electricity being generated by the power company, and that being consumed by the population on the grid. When there is an increase in usage, the generators reduce the frequency of rotation, resulting in a reduction in the ENF. When the demand dips, the generators speed up, resulting in an increase in the ENF (Cooper, 2008). These dips and troughs are only marginal, for example, one measurement showed a +/- 0.6 Hz variation over a 10-second period (Grigoras, 2007). An entire system will have the same instantaneous ENF value if taken at any of its power outputs, due to electromagnetic wave propagation, and this consistency has been proven on various occasions and across a number of different grids, including laboratories 420 miles apart (Cooper, 2008). The ENF is also power grid-specific, so in the US, there are three grids, the East Coast, the West Coast, and Texas. Two measurements from two separate power outlets taken on the East Coast (for example, New York and Washington, DC, for, say, a 30-second period), will share the same fluctuations around 60 Hz. If Washington, DC, were swapped for Las Vegas, the values would not correlate as they are on different grids. In Europe, power is controlled by the UCTE (Union for the Coordination of Transmissions of Electricity), so ENF readings taken at the power outlets within different European countries for the same period will correlate. The UK has its own network run by the National Grid company, as do some of the Scandinavian countries (Brixen, 2007a).

So how does this all relate to audio forensics? In research published in 2003 by Romanian audio expert Dr Catalin Grigoras, it was shown that ENF is captured as a signal component by audio recording equipment if the device is mains-powered and lacking an ideal voltage regulator and perfect shielding (Grigoras, 2002). It was also discovered that it can be captured by devices that are recording within close vicinity of mains power due to electromagnetic radiation fields. In tests conducted between the summer of 1998 and autumn of 2000, it was found that there was no periodicity in the fluctuations, and as such, they can be considered to be random.

Application

When these factors are combined, the potential exists for comparisons of the hum captured within an audio recording to be compared against that output by a mains outlet, to provide an opinion on the time at which the recording was captured. Even when no fundamental ENF component has been captured (such as in cases where it has been captured through electromagnetic radiation), it has been shown that up to and including the 3rd harmonic of such can be used for analysis. Harmonics which exceed this contain too much masking from acoustic sources for the component to be of any use. The capture of an ENF component is dependent on a number of factors, including microphone type, the power supply used, location of the microphone in relation to the source (for acoustic capture), compression, and capture bandwidth. With regards to battery-powered recording devices which are not connected to the mains (including mobile phones), there are essentially two methods by which an ENF component is captured within an audio recording:

- electromagnetic fields (Brixen, 2007b);
- acoustic mains hum, such as lights, power adaptors, and fans (Fechner and Kirchner, 2014).

A device just being in the vicinity of electric cables does not necessarily ensure capture of the component. Further to this, as the ENF is power grid-specific (such as the US East Coast, West Coast, and Texas), an opinion can also be given in relation to the approximate location in which the recording was captured. It can also be used to detect local editing, for example, if the ENF fluctuations of a 2-minute recording match the database, but 5 seconds in the centre do not, the edit points can be surmised to be located either side of the 5-second section.

For all of these analyses to be possible, the micro fluctuations of the ENF component must be known for comparisons to be made. A database must, therefore, be created that captures the output of the AC outlet 24 hours a day, 7 days a week, 365 days a year (Sanders, 2008). This can be performed using a computer, a sound interface, and a transformer (to reduce the mains output to a safe level, thus preventing damage to the sound interface and clipping of the signal). It is recommended that the sound interface has a minimum signal to noise level of -94 dB to ensure that noise does not reduce the precision of the ENF component. The audio should be captured in an uncompressed format such as linear PCM (although research has shown MP3 capture also to be suitable), and be connected to a UPS (uninterruptible power supply) to mitigate against unexpected power failure and subsequent loss of data (Grigoras et al., 2011). Another obvious step to prevent data loss is to perform a regular back-up of the database. It is also possible to use a smaller device to reduce the resources used by a full-size computer (Zjalic et al., 2017).

For analysis, the evidence and database signals require some preparation to ensure they are optimised to provide the most accurate and reliable results. A standard method for this is as follows (Cuccovillo and Aichroth, 2017):

1. Downsample the recording to 2.4 x F_0 Hz, with F_0 being the ENF centre frequency of interest (50 or 60 Hz).
2. Apply a steep bandpass filter around F_0, with a passband of 1 ~ 2 Hz.
3. Compute the FFT using a frame length of approximately 25 seconds (the duration necessary for a frequency resolution of 40 MHz).
4. Extract the ENF trajectory by means of peak tracking nearby F_0 in conjunction with an interpolation scheme.

A comparison of the extracted ENF components of both the evidence and database recordings can then be performed (Figure 11.9). This comparison is a clear case of the time-frequency resolution problem within signal analysis, as both the temporal and frequency information are

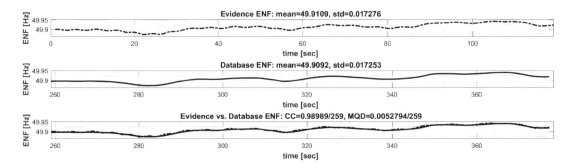

Figure 11.9 ENF comparison. Upper: Evidence ENF component, Middle: Database ENF component, Bottom: Cross-correlation calculation between the signals ENF components

Source: Created using Mathworks Matlab.

both highly informative when performing analysis in relation to the ENF signal component. There is always a trade-off between frequency and time, and as demonstrated in earlier chapters, it is not possible to represent a signal in both a high frequency and high time resolution. Exemplar and evidence are both processed with the same settings to mitigate issues in relation to this problem.

Since the initial research was published by Grigoras, a large amount of research has been performed to improve analyses in relation to audio forensics. There are essentially two types. In cases where there is information available with regards to the purported capture device (thus, a verification task), comparisons can be performed manually, using visual analysis of the spectrogram. Vertical magnification can be applied and comparisons made between the evidence ENF and a recording from the same grid and period from a database.

If there is no information available in relation to the recording, then the task of finding the exact period which matches the evidence from a number of extensive databases is nearly impossible, so automated search methods are therefore preferable (Cooper, 2008).

The following systems essentially form the foundation of all automated analyses, although some include additional steps and differ in the information used from each recording for comparison.

1   *Zero crossings (time-domain).* The number of zero crossings within a given period can be compared between the evidence ENF and a recording from a database using cross-correlation (Huijbregtse and Geradts, 2009).
2   *Phase estimation.* Fourier Analysis is used to estimate the phase by using the DFT of the first-order derivative. Traces of edit points are then exposed as abrupt changes to the phase. A feature list can also be used to quantify the discontinues for automated detection (Nicolalde Rodriguez et al., 2010).
3   *Maximum magnitude (frequency domain).* A series of power spectrum calculations are made from consecutive windows, and the maximum magnitude of each determined. These are then used to compare the evidence ENF and a recording from a database (Archer, 2012).

MDCT coefficients

Theory

MDCT is the most common approach for perceptual audio encoding, used by MP3, WMA, Vorbis, MPEG-2 AAC, and MPEG-4 AAC, as it provides energy compaction similar to the Discrete Cosine Transform used for JPEG image compression. This is achieved through critical sampling (selectively sampling specific regions), reduction of the block effect (through a 50 per cent time-domain window overlap to smooth the block boundaries), and flexible window switching to reduce other artefacts such as pre-echo in the case of insufficient time resolutions. It is a Fourier Related Transform with a cosine basis function (Wang and Vilermo, 2002).

The MDCT process appears third in the sequence used for perceptual encoding of digital audio summarised below (Luo et al., 2014).

1   Recording divided into frames (in this example MP3 is used, which has a long frame size of 1152 samples).
2   Application of a filter bank to split each frame into 32 frequency bands of equal bandwidths.
3   Application of MDCT on each band to obtain 18 coefficients per band (1152/32 = 36). As the MDCT function provides half as many outputs as inputs, this results in 576 coefficients per frame.
4   The psychoacoustic model then analyses coefficients to calculate masking thresholds.
5   Quantisation is performed using masking thresholds from the previous stage to remove the less audible components.
6   The quantised coefficients are then encoded using lossless coding into a bit-stream.

The MDCT coefficient is used in addition to the filter bank (a combination known as a hybrid filter bank), as when used in isolation, the filter bank is prone to artefacts due to its imperfect reconstruction. The above process is simply reversed when a signal is decoded, and as such, an inverse MDCT function is used. Another reason for the addition of the MDCT stage is based on the fact that the filter banks are signal-specific and time-varying, allowing adjustment of the lengths of windows dependent on the properties of the signal. Before the MDCT stage, each sub-band signal is windowed to reduce artefacts at the edges of the time-limited signal. There are various windows available, and the type of window will be determined by the content analysis performed by the psychoacoustic model.

For instance, in the case of MP3, two window lengths are used (Raissi, 2002). If there is little difference in the sub-band signal between the current and previous frame (consistent with a stationary passage), a long window is applied to maximise coding efficiency (or coding gain) and achieve good channel separation. If there is a significant difference between the two, then a short time window is applied to improve the time resolution, consisting of three short overlapped windows. This is an extremely effective technique for reducing pre-echo in transient signals as the shortened frame lengths lessen the duration of pre-echoes, resulting in them going undetected by the human auditory system.

Application

When audio is decoded, we are presented with audio samples. Repeating the previously documented encoding steps on these audio samples, but stopping once the MDCT stage is complete can provide us with the MDCT coefficients, but with a degree of error due to the quantisation which the decoded samples will have undergone (Yang et al., 2012). In essence, the analysis requires the following stages:

1   The signal is split into short time windows (1152 samples per frame (M), with a 50 per cent overlap).
2   The windows are divided into 32 sub-bands.
3   MDCT function is applied (resulting in 576 coefficients per frame).
4   The coefficients are normalised.
5   The normalised coefficients are plotted as a map of 576 x M values.

As different encoders use different filters, frame lengths, and masking thresholds, the artefacts of lossy compression are visible as vertical bars with a distribution dependent on the encoder used. It also allows the differing areas of bit allocation to be visible in both the temporal and spectral domains. This method can be used on WAVE PCM recordings to reveal the presence or absence of previous MDCT lossy compression (Figures 11.10 and 11.11), traces of upsampling, different types of lossy compression, and traces of other MDCT algorithms including AAC, MP3, and WMA (Grigoras and Smith, 2019).

Another method proposed by Luo et al. (2014) uses MDCT coefficients to expose the quantisation effects in different frequency regions by analysing statistics from the corresponding coefficients to identify compression history. Once the MDCT coefficients have been extracted, features are derived to allow the quantisation effects at different compression schemes and bit rates to be measured. The first feature exposes the fact that most lossy compression schemes implement a low pass filter, resulting in high-frequency coefficients where each frame is quantised to zero. The average number of MDCT coefficients with a value of zero per frame will, therefore, be more substantial than uncompressed recordings, allowing a judgement to be made as to whether a WAVE recording was previously compressed. To implement the method, once the MDCT coefficients have been extracted, the average number of exact zero values per frame is calculated by dividing the total number of zero values by the total number of frames. By applying a threshold, original audio clips can be detected from their corresponding compressed versions when a random bit rate and compression scheme are applied to an accuracy of up to 96 per cent.

A second method is to analyse the distribution of the MDCT coefficients for previously compressed WAVE PCM recordings. For this to be performed, the MDCT coefficients are averaged for all frames, and absolute values of the resulting mean values are calculated, resulting in 576 absolute values per recording. When plotted, this shows that the energy of different frequency components decreases from low frequency to high frequency, where the lower values pertain to high-frequency coefficients, revealing the differing degrees of quantification for different spectral regions. This can then be used to opine on the bit rate of the original audio. The methods can then

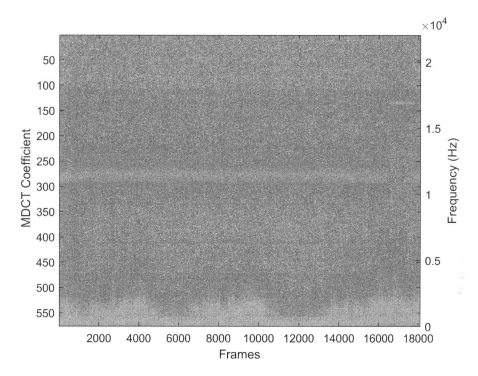

Figure 11.10 MDCT coefficient map of an uncompressed signal. Note the random distribution.
Source: Created using Mathworks Matlab.

be implemented into a feature vector (a vector which combines a number of features at varying weights) to identify the compression history of a WAV PCM signal.

Butt splice detection

Theory

The term 'butt splicing' refers to the 'butting' together of two audio regions. In a natural acoustic environment, the differences in quantisation levels between consecutive samples are expected to be relatively small due to the attack and decay portions of naturally occurring sounds, combined with the rate at which audio is sampled. When edits are performed, if no consideration is made for the differences between the quantisation level of samples, this can result in a large disparity between the quantisation levels of consecutive samples at the point at which the butts meet (Figure 11.12). This can cause an abrupt discontinuity, which may result in an audible click depending on the magnitude of the discontinuity and the amount of perceptual masking at this point. Even when attempts are made to edit at zero crossings (the point at which consecutive

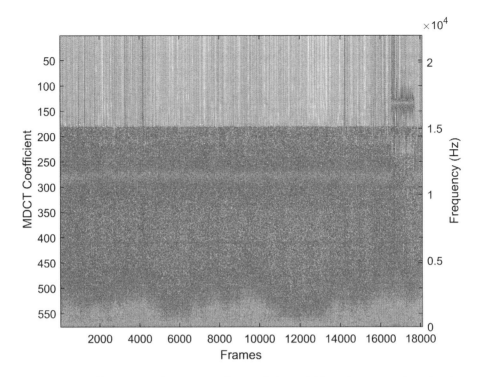

Figure 11.11 MDCT coefficient map of a perceptually encoded signal. Note the non-random distribution in the upper regions.
Source: Created using Mathworks Matlab.

samples exist either side of the central axis and thus 'cross' 'zero') between sections of equal amplitude, this does not guarantee that a click will not be audible.

Application

There are a number of methods available to detect butt-splicing, the simplest of which is to apply manual vertical magnifications to the local region within the waveform representation of an audio recording. Any traces can be better discriminated by calculating the first- or second-order derivative of the audio data, and large derivatives from the mean analysed further. Differencing (such as calculating a derivative) is most commonly applied to data to remove trends or make non-stationary time series signals stationary, similar to the technique used for high pass filtering.

When there is a specific area of concern within a recording (which may have been determined through critical listening or spectrogram analysis), manual analysis of the region is relatively straightforward. When a potential butt splice has not been identified through other analyses, this approach is much less practical and is analogous to looking for a needle in a haystack.

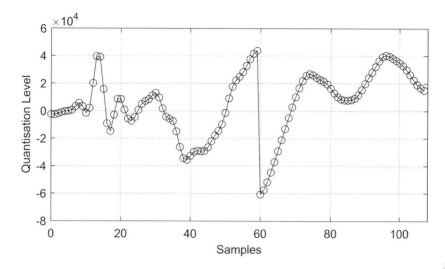

Figure 11.12 The magnified waveform of a signal containing a butt-splice between samples number 59 and 60
Source: Created using Mathworks Matlab.

Automated methods of butt splice analysis which require no *a priori* knowledge are therefore essential. Cooper, formerly of the MET Police's audio laboratory, proposed a method optimised for non-perceptual encoding and detection of edits which are not discernible through critical listening (Cooper, 2010). This is performed by first modelling a time-domain discontinuity caused by butt-splicing through second-order differencing of consecutive samples. This is then aligned to the start of the section to be searched. The cross-correlation between the audio samples and the model is then calculated and stored (principally between consecutive sample points), and the model moves forward one sample and repeats the process until the entire region has been analysed. Points of high cross-correlation are then examined further using the aforementioned manual visual techniques. It was found that a model composed of 80 samples produces the best results, as one with too many samples increases both the mean and the variance due to the disparity in content between regions, resulting in false positives for signals of poor SNR. Too few samples and there is little discrimination between the discontinuities and the rest of the audio.

Spectrogram

Theory

Spectrum analysis refers to the analysis of audio within the frequency domain, for which there are various methods, all of which require the application of FFT in some capacity. Although useful, visualisation in the time domain provides only a limited perspective as there is so much information contained within an audio recording which does not pertain solely to the timing and magnitude of each sample. Representation in the frequency domain is, therefore, a critical stage of any analysis. The most commonly known and easily accessible method of frequency representation is the spectrogram. This allows an audio recording to be visible in the frequency, time and magnitude domains simultaneously, providing an overview of its composition. Viewing the signal in the time and frequency domain in a single window allows the relationship between the frequencies relative to temporal regions of a recording to be reviewed, some of which may not be audible.

Application

As perception is highly subjective, the data visible within a spectrogram lends itself to providing an objective demonstration of any artefacts. Take a recording of an interview which contains a 50 Hz hum which may not be discernible, and so was not documented during critical listening. When viewed on a spectrogram, a clear 50 Hz tone throughout the majority of the recording can be identified, except for a short break in the middle. A hum of this frequency is caused by the recording picking up the AC power frequency (on mains-powered devices, 50 Hz in the UK, 60 Hz in the US), or interference from other electrical devices within the magnetic field of the recording device. If the recording is known to have been captured using a mains-powered device in a room in which there are no other electrical items, the region in which this tone is not visible may be indicative of the said section not being part of the original recording.

Another area for consideration is the frequency composition of speech. Speech synthesis algorithms result in high-frequency harmonics which extend beyond the frequencies produced by the human voice and can also cause harmonics which bridge gaps between words that are highly correlated with the lower formants of speech. These traces may not be audible, and would not be visible when viewing the recording in the time or frequency domains in isolation. As the artefacts pertain to both time and frequency, they would only be visible within the spectrogram.

When performing spectrogram analysis, considerations should be made as to the optimal settings of both the input window length and the number of FFT sample points. These settings should be based on the nature of the recording, the focus of the analysis, and FFT theory (Koenig, 1990). As covered in earlier chapters, there is a trade-off between frequency and time resolution when performing this type of analysis, and caution should be exercised when drawing comparisons between the spectrogram data pertaining to different recordings due to potential differences in resolution. The most straightforward approach to finding the optimal settings with regards to a specific recording is as follows:

1   Determine the type of signal which is being analysed. Is it a signal with a lot of transients which requires a relatively high temporal resolution, or one containing a stationary signal such as a constant hum? If it is the former, it would need shorter frame lengths to ensure the transient is not lost among the other samples. If it is the latter, it would require a longer frame to provide a higher resolution in the frequency domain to achieve an accurate representation of the hum.

2   Determine the FFT window size. This can be achieved through trial and error (for example, starting at 1024 points and adjusting until the pertinent features become visible) or through the application of the following theory:

   $N = Fs/\text{Spatial resolution}$

where N is the number of FFT sample points.

By way of an example, if it is calculated that the length of the smallest component being analysed is 50 samples in the temporal domain, and the sampling rate is 44.1 kHz:

N = 44100/50

N = 882 points

This should then be rounded to the nearest power two (the reasons for this were covered in Chapter 5), which in this case would be 1024 points.

3   A Hamming window with a 50 per cent overlap is a good starting point as it reduces edge artefacts without an extreme reduction in the data at the extremes. It, therefore, ensures excellent coverage of all signal types within the recording.

In summary, spectrogram analysis is the primary method for visual analysis of elements within a recording which pertain to the spectrum as it contains frequency, amplitude, and temporal information. The drawback is the subjective nature of its interpretation, so, as always, cross-correlation against other analyses is of paramount importance.

Power spectral density

Theory

As PSD is covered extensively in Chapter 5, the reader is encouraged to reread it, to gain a greater understanding of the theory before learning how it can be applied for authentication purposes.

Application

For global analysis, the PSD can reveal traces of lossy compression through peaks and troughs within the distribution of the spectrum caused by perceptual encoding. The PSD of the first derivatives can also be sorted to provide more detailed results of the changes in the energy across the system, and plotted as the 'Differentiated Sorted Spectrum.'

For local analysis, the power trajectory of a signal can be calculated over short time frames. This can provide information in relation to signal inconsistencies in the time domain (Grigoras et al., 2012), for example, in regions where the energy for a specific section may appear higher or lower than the surrounding area. This could be indicative of signal loss (which may be caused by insertion of silence, mechanical failure, or compression), or a splice (where there is a distinct inconsistency between the power of one section and the rest of the recording).

Long-term average spectrum (LTAS)

Theory

The LTAS is a specific type of PSD representation. It is a broad term pertaining to the calculation of the average short-term Fourier magnitude spectra, and, thus, there is no standard, so it can be performed in a number of ways. Welch (1967) proposed a short-term averaging method in which the signal is divided into overlapped frames before the power spectra of each frame is computed, and the average of the spectrum then plotted. Another method is to apply FFT to the entire signal, then use a windowing function and zero pad to the next power two. As the spectrum length varies depending on the input, power spectrum points are reduced by averaging neighbouring frequency bins. The results are most commonly transformed into a log representation and sometimes normalised with the strongest component represented by 0 dB (Kinnunen et al., 2006).

The primary use of the LTAS is to provide a holistic overview of the frequency composition of the signal, as it visualises the frequency distribution of the sound energy. Another technique proposed by Grigoras (2010) is to calculate both the minimum and maximum frequencies contained within each frame, which, once plotted, show the upper and lower limits of the magnitude of a signal at specific frequencies.

The Nyquist-Shannon theory states that the highest component of a signal should be half the rate at which it is sampled, and although metadata can provide this information, not only can it be easily manipulated, it does not offer any information as to the distribution of data across the frequency range occupied. LTAS is, therefore, a quick and reliable method of determining this.

One of the most significant advantages of LTAS is that if the signal is long enough, individual differences such as the articulation of speech, accents, and the content do not affect the mean value, as they are represented by an average, thereby reducing the effects of variation (Kitzing, 1986). It is, therefore, the ideal method when comparing exemplar recordings which were created to match the general content of an exhibit, without the characteristics of voices, such as the accent and general articulation impacting the result.

One consideration to be made is the inclusion of pauses and silence when calculating the LTAS, as they can have an impact on the result. If an evidence recording contains a number of regions of silence, a threshold can be applied to only include windows which contain data of a certain amplitude or above, essentially normalising the calculation so reliable comparisons can be made against other recordings.

Application

LTAS can be used to provide information in relation to several factors for use in authentication, including but not limited to: the acoustic environment, acoustic sources, recording equipment characteristics, transmission channel characteristics, other signals picked up by the equipment or transmission channel, traces of filtering, downsampling, and lossy compression.

A histogram of each energy level can also be plotted to provide a second visual perspective which may display information not visible in the standard LTAS representation.

During an authentication examination, the first comparison should be between the sampling frequency of the recording and the range of frequencies covered within the LTAS. The range would be expected to cover from 0 Hz–Fs/2, and any disparity between these points would instigate further analysis to discover the reasons why. Second, the plot can be analysed to determine the energy distribution across the entire bandwidth. A recording captured at 8 kHz (a bandwidth of 0–4 kHz) which has then been encoded at a higher sample rate (for instance, 44.1 kHz) will contain only negligible amounts of data above 4 kHz, as a low pass anti-aliasing filter would have been applied during the initial capture, removing any data above that point. Any additional data will come from the re-encoding of the data and the associated anti-aliasing filter. Another consideration is the distribution of the energy of perceptually encoded signals. During encoding,

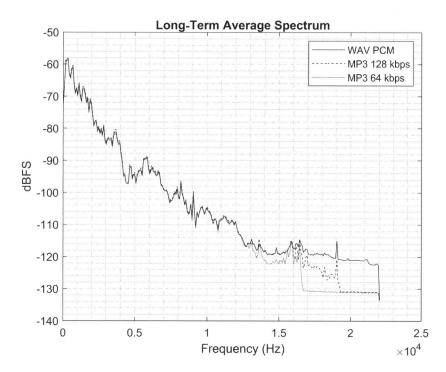

Figure 11.13 Long-term average spectrum of a signal encoded in different formats
Source: Created using Mathworks Matlab.

a filter bank is applied, followed by quantification of the sample values based on a perceptual model. In doing so, the upper frequencies are reduced in magnitude as it is an area in which the human auditory system is least sensitive. This generally becomes visible through a distinctive drop-off, the cut-off frequency and slope of which are determined by the codec and sample rate used (Figure 11.13).

When drawing conclusions, care must be taken as two encoding processes may result in similar representations. For example, although a 44.1 kHz signal may contain data which extends up to 16 kHz, with a visible drop-off in data above that, this does not necessarily mean the original signal was 32 kHz and was saved out of higher sample rate software.

Long-term average sorted spectrum

Theory

The long-term average sorted spectrum (LTASS) is the sorting of the values from the LTAS in descending order of magnitude, and as with LTAS, the mean, maximum, and minimum averages (referred to as M3) can be plotted to achieve an overall visualisation of the frequency content (Grigoras, 2010).

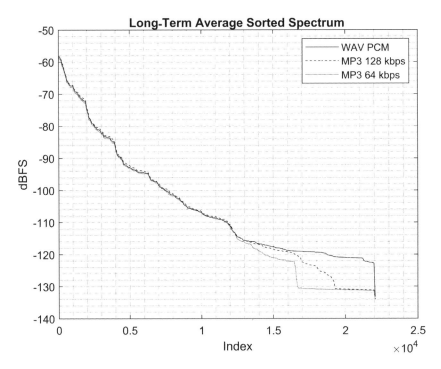

Figure 11.14 Long-term average sorted spectrum of a signal encoded in different formats. Note the differences in the magnitude in the upper regions.

Source: Created using Mathworks Matlab.

Application

The LTASS plot is much smoother than LTAS as the bins are sorted, and so the visibility of variations caused by the capture content and environment is reduced. This, in turn, aids in drawing conclusions as to the frequency response of the capture device.

To expose traces of lossy encoding or data spliced from other recordings, the first derivative can be calculated and then plotted (Grigoras et al., 2012), resulting in the presence of a small bump before a steep drop-off for low to average bit rate lossy compressed recordings. It also provides a more stable representation of the previously mentioned perceptually encoding drop-off, and facilitates the measurement of the cut off frequency by calculating the largest derivate between frames (Grigoras and Geradts, 2019). This can then be used to differentiate between encoders, as different lossy compression encoders begin to apply zero quantification at different frequencies (Figure 11.14).

When the LTASS is plotted, the horizontal axis will now represent an index rather than frequencies, as the frequencies have been sorted in descending order, so the energy at each horizontal point is not representative of the frequency distribution at that point. Care must, therefore, be taken in the interpretation of the visual results of this method.

Compression level analysis

Theory

When digital audio recordings are stored in an uncompressed format, all data is represented at the highest resolution possible, regardless of its magnitude or frequency. When recordings are compressed using perceptual techniques, the similarities between regions of data are used, and interpolation is performed to reduce the number of bits required for storage. This process results in a distinct periodicity between neighbouring samples in the time domain, as opposed to a random correlation between samples with uncompressed recordings, consistent with the acoustic sounds which it captured.

This correlation between areas of compressed data means the rate of change across samples is slower than it is for uncompressed, random signals, and can be measured by calculating the second derivative (Grigoras et al., 2012).

Application

As it applies to authentication, the above theory can be used to determine if a recording has been perceptually encoded, as the correlation between data will not be random as they would be for uncompressed recordings. As compressed audio uses frames, a measurement can be based on the rate of change between frames rather than samples. This can be achieved by first splitting the signal into frames before calculating the second derivative. The result can then be plotted, and the rate of change determined. It can also be visualised in the frequency domain through conversion using DFFT (Grigoras, 2010).

A key advantage of this technique is its robustness to transcoding. For example, a recording which was initially lossy compressed and then converted to an uncompressed format would retain the ordered effects of interpolation. Re-encoding a recording captured to an MP3 format a second time (creating a second-generation MP3 recording) would result in further correlation and further traces. This can assist the examiner in determining whether the recording has been transcoded since capture, providing exemplar recordings can be created from the same device to allow comparisons to be performed. Its use is limited if the file provided is perceptually encoded and the purported capture device is unknown, as compression has taken place and there would be nothing for the results to be compared against to determine whether it has been encoded more than once.

DC bias

Theory

The direct current (DC) component of a signal is defined as the mean of the signal amplitude within a single period. As sound can be decomposed into sine and cosine waves containing mirrored peaks and troughs, it stands to reason that if the amplitude values are shifted to a bipolar representation, the mean value would be zero (in a perfect system).

In reality, the capture of digital audio involves the use of direct current, regardless of whether the device is battery-powered or mains outlet-powered because AC power is converted to DC

power by a transducer before running through the circuitry of a device. As the electrical components have slight imperfections, the potential to disrupt the perfect balance exists if DC voltage
from the said components is captured with the recorded signal. This occurrence results in a DC
bias. Bias is defined by the American National Standards Institute (2019) as:

- a systematic deviation of a value from a reference value;
- the amount by which the average of a set of values departs from a reference value.

In professional and consumer electronics, the majority of this DC bias is removed by a blocking
capacitor before the analogue to digital conversion stage, but not completely. The reference value
for audio recordings is considered to be zero, as that is the ideal response of a system, and in a
perfect world, a recording is therefore expected to be centred on the x-axis when the waveform
is analysed in the time domain. If there is a positive DC bias, the result is a positive vertical shift
of the centreline above the x-axis. If there is a negative DC bias, there is a negative vertical shift of
the centreline below the x-axis. The extent of the shift is based on the amount by which the value
departs from the reference value of zero.

The DC bias is essentially the mean of all quantisation levels within a digital audio signal. It
can be calculated in quantisation values using Equation (11.1), where T represents the total number of samples, and QL represents the quantisation levels.

$$DC(QL) = \left(\frac{1}{T}\right) x \int_{-T/2}^{T/2} f(x)dx \tag{11.1}$$

It can also be calculated as a percentage, using Equation (11.2) for positive mean and Equation
(11.3) for negative mean values:

$$DC(\%) = DC(QL) x \left(\frac{100}{2(n-1)-1}\right), DC(QL) > 0 \tag{11.2}$$

$$DC(\%) = DC(QL) x \left(\frac{100}{2(n-1)}\right), DC(QL) < 0 \tag{11.3}$$

And as dB using Equations (11.4) and (11.5):

$$DC(dB) = 20log\left(\frac{DC(QL)}{2(n-1)-1}\right), DC(QL) > 0 \tag{11.4}$$

$$DC(dB) = 20log\left(\frac{DC(QL)}{2(n-1)}\right), DC(QL) > 0 \tag{11.5}$$

Although percentages and dB measurements allow comparisons between systems of different bit-depths, quantisation levels are the recommended form of measurement as they have been shown to be the most precise (Koenig et al., 2012).

Application

The occurrence of a DC bias is taken advantage of within digital forensics for two reasons. First, the cumulative effect of the electronics results in a DC bias which is consistent between recordings captured with the same device, providing the general acoustic content is similar, and the format, specifications, and microphone are the same. This allows comparisons to be made between the DC bias of an evidence recording against exemplar recordings from the purported capture device to determine if they are consistent. Conclusions can subsequently be drawn as to whether the exhibit may have been captured with the said device.

Further to this, even though specific models of recording devices are mass-produced in a controlled assembly-line manner using the same components for each device, there are still slight disparities between DC biases of matching recording devices due to minute differences in these components. When compared against recording devices manufactured by other companies, or different models created by the same company, the differences can be more substantial, as they will use various components, unique combinations of components, and differing circuitry.

With that being said, this technique cannot be used for identification of a device as the offset cannot be considered to be highly individualistic. It should, therefore, only be used as a method to exclude the possibility that the exhibit was captured with the purported device, providing the device is available to create exemplar recordings (Fuller, 2013). An extension of this would be when there are a number of exhibits, and an opinion is sought as to whether they all originate from the same device by comparing the offsets from each exhibit against one another.

DC bias analysis should also only be applied to recordings over 1 minute in length, as research has shown the shorter the duration of the recording, the less reliable the results, as the average is calculated from fewer samples. It has also been found that the DC bias is affected to a small extent by lossy compression, and the lower the compression quality, the more impact it has on the DC bias. This should, therefore, be taken into consideration when drawing any conclusions.

As an entire input signal will be processed using the same electronics, it is expected that there will be a low disparity of the DC bias between all regions of the recording (SWGDE, 2018). By splitting the signal into frames and calculating the DC bias per frame, conclusions can be drawn as to whether the entire recording was captured using the same device (Figure 11.15). For example, a recording which consists of a DC offset which is generally between -0.05 per cent and +0.05 per cent, but contains a short section in which the offset is +2.00 per cent may be an indication that a section has been spliced into the recording from a different source. The local DC offset calculation must be performed over relatively moderate frame sizes (of at least 1 minute) to ensure an accurate average is calculated.

Zero padding

Theory

The term zero padding refers to the appending of zero value samples to a signal. It is generally used to increase the length of a recording for FFT analysis with an improved resolution (Donnelly

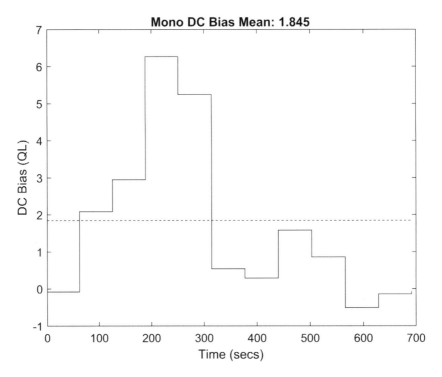

Figure 11.15 Staircase plot of a signal's DC bias in 1-second frames
Source: Created using Mathworks Matlab.

and Rust, 2005), or to ensure a signal complies with the format specifications for which it is being captured (Allamanche et al., 1999; Sripada, 2006). The addition of zeros for FFT representation would only occur post-capture during analysis, so it is the latter which is of interest to forensics.

As discussed in previous chapters, during perceptual audio encoding, the signal is divided into frames, the length of which is defined by the specification of the format used. For example, during AAC encoding, a transform is applied to consecutive sets of 2048 audio samples (2 x 1024 sample frames). As two frames are required, there is a prerequisite for an addition of 1024 samples before the first 'true' audio sample to allow the first true frame to be read. Thus, a single frame of 1024 zero-value samples is added to the beginning of the recording during encoding.

In addition to this, there is also an 'encoder delay.' This is the delay incurred during encoding to allow the production of properly formed samples, and adds further zero value samples to the beginning of the recording, known as 'priming samples.' There is then a 'decoder delay,' which is applied to allow the proper decoding of the samples (Figure 11.16). As the signal is also encoded in frames (or packets), additional samples are also appended to the end of the signal to ensure the specification is met, and the frames are all of the correct size (Apple Inc., 2016). For instance, if there were 5121001 samples in a signal, 5000 frames of 1024 samples would be created, with

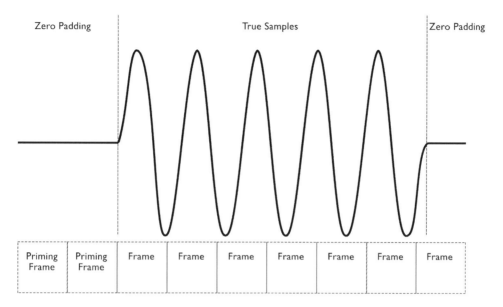

Zero Padding   True Samples   Zero Padding

| Priming Frame | Priming Frame | Frame | Frame | Frame | Frame | Frame | Frame | Frame |

Figure 11.16 Perceptual encoding zero-padding overview

the remaining number of samples (1001) being too small to form a frame. Thus, 23 zero-value samples would, therefore, be added to the signal to meet the required frame size according to the specification. In light of the above, upon decoding, the first true audio sample will be offset by a certain number of zero value samples.

For the playback of standalone audio, there are no issues caused by the additional samples which will impact the perceived output, but for video, it is a different story. As there are additional samples, the decoder needs to know the number of samples to ensure the video and audio streams are in synchronisation. These can then be removed once playback begins or when the user navigates to another section of the recording. For this reason, companies use standardised amounts of zero padding for their encoders, allowing decoders to be developed that can accurately remove the zero samples values at the beginning of a recording. This does not apply to the zero samples at the end of the recording, as (1) they do not affect the synchronisation as they occur once the audio has finished, and (2) they are dependent on the number of true audio samples within the recording.

It must also be considered that when an audio editing suite is used to open a recording, the decoder processes the audio and presents it in an LPCM format. If this recording is then saved out of the software in a compressed format, the software's encoder will consider the first sample to be the first 'true' audio sample, thus repeating the process and padding the beginning of the recording even further. It is not possible to predict the padding at the end of the recording as this is dependent on the number of samples required to meet the frame specifications.

As WAVE PCM is stored as a representation of the true bit-stream and does not undergo any form of division into frames, no padding is required. If a perceptually encoded recording is saved

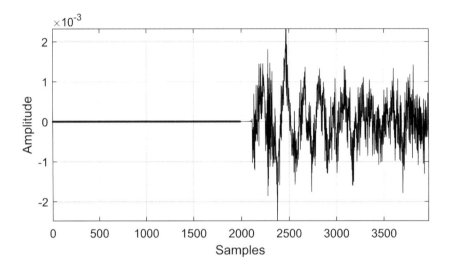

Figure 11.17 Casework example of the zero-padded component of a perceptually encoded signal
Source: Created using Mathworks Matlab.

out of software in a WAVE PCM format, it will therefore not pad the recording, and the number of zero values samples previously padded to the beginning and end of the recording will be maintained.

Application

This process has several applications to forensic audio authentication. First, as specific codecs pad a set number of zero value samples to the beginning of a recording, comparisons can be made against exemplar recordings from the purported capture device, or against a database of exemplar recordings from various codecs. This can provide information as to the originality of the recording, but as the number of appended zero value samples is not specific to a device, it does not provide information in relation to the specific source device, only a specific codec (Berman, 2015). For example, one well-known company uses 1984 zero-padded samples at the beginning of their audio recordings, so if a recording is purported to have come from a device created by the manufacturer and does, in fact, have 1984 zero value samples at the beginning of the recording (Figure 11.17), then it is 'consistent' with having come from that device, at least with regards to zero padding.

Second, as we know that WAVE PCM recordings do not require padding, it would be suspicious if a recording did contain consecutive zero value samples at the beginning or end of a recording, and may be a sign that the recording was initially captured in a perceptually encoded format and has since been converted to WAVE PCM. Following on from the example previously used, if the recording has 1984 zero value samples at the beginning of the recording, conclusions could be drawn in relation to the recording's provenance, as we know the encoder that uses this precise number.

Third, the information can also be used to determine the generation of the recording, as we know that a recording which is perceptually encoded twice will repeat the zero-padding process. If this process is repeated once more, further zeros will be added, and so on, and so forth. If a database of re-encoded exemplar recordings is created, comparisons can be made and conclusions drawn as to the generation of a recording.

Phase

Theory

Phase defines the relationship between two or more waveforms and is measured in degrees. Although the degree of phase shift is dependent on the temporal difference between the waveforms, it is also influenced by the frequency of the content. If there is a delay in a complex waveform with regards to another, each frequency of which the signal is composed will have its own degree of shift in relation to the other frequencies. The use of phase as a measurement is generally applied for the comparison of two identical waveforms or captures of the same event from different microphones. The result can be classified as 'phase-coherent' or 'in-phase' when there is no phase shift, or 'out of phase' when there is (Izhaki, 2008).

When dealing with true stereo signals, the cause of phase is inter-aural time and level differences between the two microphones. Take a single short transient burst, such as a gunshot. As the sound has travelled a greater distance to the furthest microphone, there will be a slight delay and magnitude drop in comparison to the closer microphone. Now imagine the same signal captured using a single microphone, which is then converted to double mono. Drawing conclusions from critical listening as to whether the signal is true stereo or double mono may not be possible due to the small magnitude of the differences in both time and amplitude (dependent on the spacing between the microphones), and this would also be the subjective opinion of the examiner. It may be possible to draw conclusions from viewing the magnified signal from each channel in the time domain, but this would be time-consuming if performed for an entire recording.

Application

By determining the classification of the phase ('phase-coherent' or 'out of phase'), the true spatial composition of a stereo signal can be determined. At its most simple this can be achieved by subtracting one channel from another in the time domain (essentially inverting the phase of one signal in relation to the other). Although the reported specifications may document whether the exhibit is mono or stereo, this will not provide any further detail with regards to the content of each stereo channel. By comparing the results of a phase inversion against exemplar recordings from the purported capture device, and the capabilities of the said device, an opinion can be given as to whether the channel configurations are consistent. Imagine a case in which the purported device has two microphones and captures the sound in two separate channels. The metadata states only that the recording is 'stereo.' When one channel is subtracted from the other, the result is a signal consisting of only zero value samples (Figures 11.18 and 11.19), indicating the same

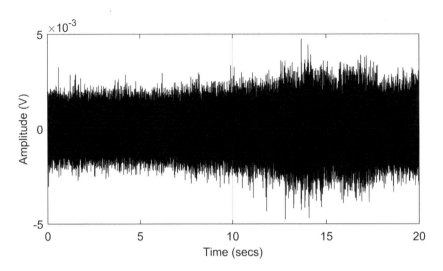

Figure 11.18 The phase difference between two channels of a stereo recording
Source: Created using Mathworks Matlab.

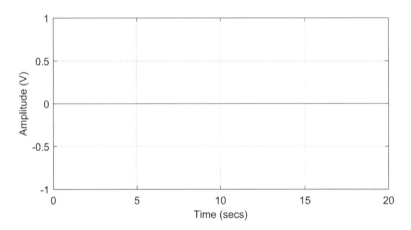

Figure 11.19 The phase difference between two channels of a double mono recording
Source: Created using Mathworks Matlab.

information exists on both channels (as subtracting a sample value from a copy of itself will equal zero). The analysis, therefore, shows the recording is, in fact, composed of two identical channels, so could not be an original recording captured using the purported device.

Aside from the global method, this analysis can also be used for local detection. Imagine that instead of containing two identical channels, the result of phase analysis is a signal with a small amplitude (indicating there is a difference between the channels, and thus it is a true stereo capture). In the middle of the phase analysis plot is a section of zero value samples, 5 seconds in duration (showing no difference between the two channels, and thus double mono). Further analysis of this area can then take place as it is plausible that the 5-second region was a monaural recording spliced into a true stereo recording. As audio cannot be coded to switch between channel configurations during the capture, a mono recording added to one which exists as stereo requires the software to double the mono recording to meet the specifications of the session.

Quantisation level

Theory

The process of quantisation is an essential component for the capture of digital audio, and, as such, every audio sample within a recording has undergone quantisation, regardless of the recording device, format, and codec used. With that being said, the actual process applied differs depending on the format, and although there are a number of bit depths available, it is rare to encounter any other than 8 or 16 bits in forensic audio due to the requirement of devices to balance quality with storage capacity.

For WAVE PCM recordings, the process of quantisation is relatively simple. Upon capture, each sample of the electrical signal (which was previously converted from acoustic energy) is rounded to the nearest level, where the bit depth determines the number of levels available. A recording with a bit depth of 8 bits will have 256 levels available to represent the amplitude of the signal. In comparison, a recording of 16 bits will have 65,536 levels available (the bit depth is a log two representation, so 2^{16}). As the input signal is continuous and sampled over tiny increments in time, it is natural for a signal to occupy consecutive quantisation levels, so there are few gaps between levels. With that being said, it may be that not every quantisation level is used, especially on recordings of lower magnitude. For example, a low magnitude recording in which the highest quantisation level used is 10000 will have empty bins beyond this if a histogram was plotted of all the used values. A histogram is a plot showing the number of occurrences of values in a sequence, or in this case the number of occurrences of each quantisation level used in a digital audio signal.

Most perceptual encoders quantise using a power law quantiser, and its application depends on the audio content. Rather than the quantisation noise for each spectral coefficient producing the same average noise as it does for a uniform quantiser, the power law quantiser distributes the distortion towards coefficients with large amplitudes where perceptual masking will be higher, thus maximising efficiency (Herre, 1999).

Application

This theory is the basis for a number of analyses proposed by Grigoras and Smith (2014) which can be applied to expose changes to a recordings bit depth or amplitude. It can also reveal traces relating to the application of a filter and splicing.

The first method presented is to plot a histogram of the signal's quantisation values. As the signal which was sampled is continuous, when the difference in quantisation levels between each consecutive sample is plotted as a histogram (where the difference between each are the bins),

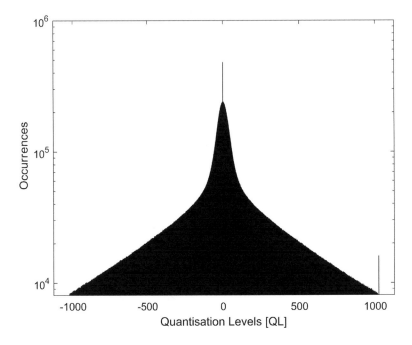

Figure 11.20 Quantisation level histogram of an uncompressed 16-bit WAVE PCM recording
Source: Created using Mathworks Matlab.

the result is a Gaussian-type distribution (Figure 11.20). This is caused by the fact that the gaps between consecutive quantisation levels are minimal due to the continuous nature of acoustic sound and the high sample rate of audio.

Just because there are a certain number of quantisation levels available, it does not necessarily mean all are being used. For example, although there may be 256 levels available within an 8-bit recording, the actual audio captured may only occupy 130 of these levels, meaning the recording content is only 6.9069 bits (calculated as the log two of 130). As it applies to WAV PCM recordings, when a recording is captured at 8 bits, but then transcoded to 16, there is an increase in quantisation levels available for the amplitude information to occupy. What was originally distributed across 256 levels is now dispersed across 65536 levels, equating to 256 levels for every level in the 8-bit signal (Figure 11.21).

When this transposition occurs, a large number of steps will exist between these values which are now unoccupied. The minimum quantisation jump between two consecutive samples will still be zero (imagine two clipped samples), but the second smallest jump will now be 256 quantisation levels rather than 1. By plotting a histogram of the difference between consecutive sample values, these gaps become visible, consistent with an increase in the bit depth of the exhibit (Figure 11.22).

A second method counts the number of unique quantisation levels and computes the next power of two, thus giving the actual bit depth of the recording. This can be performed on the

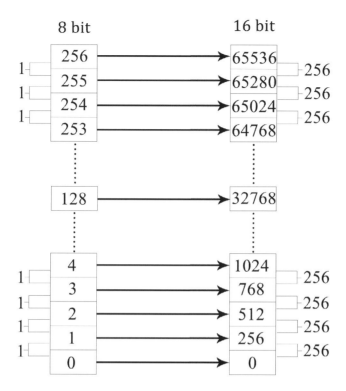

Figure 11.21 Overview of the transposition of quantisation levels during an 8-bit to 16-bit conversion

entire signal as a form of global analysis, or signal split into frames as a local analysis (a technique called quantisation level trajectory) to reveal sections of the audio which have undergone processing. For example, a native 8-bit signal which has undergone de-clipping in an audio suite and saved out at 16 bits will reveal the de-clipped sections at 16 bits against the native 8-bit signal.

As a global example, a 16-bit recording which contains only 14 bits of audio content must have been captured by either a 16- or 24-bit system, as 14 bits will not fit into 8 bits. This information can also be used to make comparisons against an exemplar recording from a purported capture device, as although some specifications may state that audio is captured at 16 bits, in reality, they often capture at a lower bit depth. By comparing results from both exemplar recordings and the exhibit, an opinion can be rendered as to the consistency of the exhibit's quantisation levels with the purported capture device.

There are some caveats and areas of caution when applying this form of analysis. First, as it relates to telephone calls (which are generally of 8 bits in nature), considerations must be made for the capture system. For example, if a recording is captured at the operator side of the conversation, and the operator's voice is captured by a 16-bit system before it is encoded for transmission via telephone, then it is likely the recording will be a composite of both the 16-bit capture

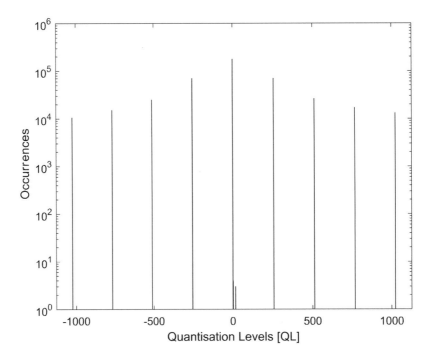

Figure 11.22 Quantisation level histogram of a WAVE PCM recording which has been converted from 8 bits to 16 bits

Source: Created using Mathworks Matlab.

of the operator and the 8-bit (converted to 16-bit) capture of the subject on the other end of the line. If the signals are captured by two independent channels, for example, the operator on the left channel and the other speaker on the right, then analysis of each channel may allow conclusions to be drawn. For mono recordings of this description, a local analysis may show differences between the bit depth of each voice, depending on the capture method.

For telephone recordings which are intercepted, the signal would be expected to be 8 bits. If the software used to capture the recordings is 16 bits, the recording will show signs in relation to this transcoding, but there would be no reason why a system only designed to intercept telephone calls would capture in 16 bits, as this would mean an 8-bit level of redundancy.

Bit depth-focused techniques have also been shown to be effective in the analysis of perceptually encoded audio recordings downloaded from the internet, but as these types of recordings use non-linear quantisation levels, the interpretation of the same analysis must be re-evaluated, and considerations must be made on a case-by-case basis due to the number of variables. While there are limited permutations for WAVE PCM recordings due to the specific range of available bit depths and the simple quantisation process, perceptually encoded formats are a little more complicated.

Take, for example, an MP3 recording which has been converted to a 16-bit PCM recording. The MP3 encoding stage will result in a non-linear distribution of quantisation levels due to the power law quantisation applied, before the data is then linearly distributed when encoded into a PCM recording. For this reason, the technique should only be applied to perceptually encoded recordings when there are exemplar recordings available for comparisons to be drawn. As with any analysis, the limitations of the technique should also be made clear in the report.

Codec analysis

Theory

Due to the limited storage capacity of mobile recording devices, combined with the requirement for large amounts of data to be stored, lossy compressed formats are extremely popular. There are over one hundred codecs in use, and as this number continues to grow each year, understanding and being able to detect the codec to encode a recording used can be very useful for authentication. Not only can codec detection provide information as to the source, but can also identify any further compression that may have taken place. For instance, a recording provided in a WAVE format may have been previously encoded using an MP3 encoder before being trimmed and exported. There are essentially two types of codecs, those explicitly used for speech, and those designed for more general use such as music and other forms of entertainment. The first consists of two subsets, based on the encoding method used. Waveform encoding pertains to PCM and ADPCM and performs coding of the waveform itself. Model-based coding, such as linear predictive coding (LPC) is based on the model of speech production and is, therefore, able to provide a higher degree of compression. Most speech codecs use code-excited linear prediction (CELP), which is based on an LPC model. Codecs used for GSM (global system for mobile communications) are designed to use the lowest bit rates due to the bandwidth limitations of wireless communication, and PSTN (Public Switched Telephone Network) has the highest bit rates compared to other speech codecs. VOIP (voice over internet protocol) bit rates are somewhere between the two.

General codecs are those that are better known due to their universal application in the real world. MP3, AAC, and WMA are all examples of this. Codecs of this type reduce redundancy through the use of time to frequency mapping, with MDCT being the most popular method. This is due to its excellent efficiency, and as such it is used in MP3, AAC, OGG (Vorbis), WMA, and AC-3 encoding. Although they all use MDCT, each format differs in their bit rate capabilities, available sample rates, and as it applies to the MDCT, the window size and type. For example:

- MP3 uses two window sizes, namely, 1152 and 384 samples in length.
- AAC uses window sizes of 2048 and 256 samples in length.
- WMA and OGG (Vorbis) have different window sizes depending on the sampling frequency but can have values of 64, 128, 256, 1024, 2048, 4096, 8192, or 16384 samples in length.

All of these factors can reveal traces in the encoded audio (and subsequently decoded) audio, regardless of the content.

Application

As a number of methods have been proposed for the detection of codecs, a few examples are provided which vary in the type of codec detected, the artefacts exposed, and the method used for classification.

Hicsonmez et al. (2013) performed research into a technique based on various distinguishing factors of the coding used by a number of codecs, including:

- the type of encoding (waveform, model-based or time-frequency mapping);
- the quality of the encoded audio;
- the complexity of the audio (based on traces such as coding delay and Millions of Instructions per Second, MIPS).

The method first trains a system to classify audio which has been encoded using different codecs and features, based on chaotic and randomness theories, before the trained model is then applied to audio recordings. The study reported results of close to 100 per cent accuracy for single and transcoded audio and results around 80 per cent for twice transcoded recordings.

Rather than use machine learning, Derrien (2019) focused on AAC codec detection using quantisation errors in the time-frequency domain. The method is based on the theory that although global quantisation noise can be similar to audio encoded using high bit rates, the structure of the related distortion is not. The encoding process used by the codec was first replicated, and the differences between the input and output rounding process are then stored. In order for this to be performed accurately, the same start time of each sample, window length, and channel configuration with which the original version was encoded must be used. By applying every possible configuration and keeping that which is the most significant measure, before using uniform sampling of the scale factors (as the possibilities are too great to iterate through), the rounding error can be estimated and conclusions drawn. The results reported no false positives and few false negatives, and detection rates of close to 100 per cent.

Draghicescu et al. (2015) propose a method for detection of recordings encoded using AMR and G.729 codecs for mobile devices based on a speaker recognition system, but by applying it to the speech encoding rather than determining a speaker's identity. It is reasoned that any channel will affect the properties of the signal for which it is transmitting, thus introducing variations into speech signals transmitted via telephony. The proposed technique extracts features from a recording for the training of a Gaussian mixture model (GMM), which is used as it takes into account the decoding artefacts with no regard for the variance of speech. A LLR (logarithmic likelihood ratio) is then applied with a threshold set for a good distinction between random similarity, and similarity pertaining to the identified codec. The results were, as with other methods, close to a 100 per cent detection rate.

A final example of a codec detection method was proposed by Luo et al. (2015), who discovered that an artefact of AMR encoded recordings is the repetition of the numerical value of adjacent samples. This introduces small horizontal lines into the waveform, which can be viewed by sorting the samples in increasing order. To prevent misinterpretation, regions of silence must first be removed. Once this pre-processing has been performed, the rate of matching adjacent samples can be calculated by dividing the number of repeat samples by the number of samples within the waveform. The rate can then be used for codec detection of AMR encoded recordings, and to determine potential regions of splicing by calculating the rate of short time frames and locating those which have a rate significantly different from the rest.

Analysis of Results Summary

To provide transparent and easy to understand results for the examiner performing the authentication, the peer reviewer, the instructing party, and trier of facts, tabulation of the results is advised. Table 11.5 is an example of a simple layout for this report.

Table 11.5 Results table example

Analysis class	Analysis type	Result summary	Decision
Container	Filename		
	Filename Extension		
	File Format		
	File Header		
	File Structure (Hex)		
	Metadata		
Content (Global)	Critical Listening		
	Waveform		
	Spectrogram		
	Differentiated Sorted Spectrum		
	Compression Level		
	LTAS		
	LTASS		
	MDCT		
	DC Offset		
	Power Spectral Density		
	Zero Padding		
	Quantisation Level/Bit Depth		
	Stereo Phase		
Content (Local)	Critical Listening		
	Waveform		
	Power Spectral Density		
	Stereo Phase		
	ENF		
	Windowed DC Offset		
	Windowed Bit Depth		
	Spectrogram		
	Butt Splice		
	MDCT		

Drawing conclusions

Once all possible analyses have been performed, overall conclusions can be drawn based on the cross-correlation of the results of each analysis, thus providing the most objective and reliable conclusions possible. Rendering an opinion based on a single analysis should not be practised under any circumstances. In an ideal world, all analyses would point in the same direction. For instance, a minimum of two global analyses that show a recording has been edited (allowing for cross-correlation between the two), and further local analyses which pinpoint the exact edit points.

Transparency is a scientist's best weapon, and this is something to embrace as an audio forensic examiner. Presenting both the limitations and results of each analysis in isolation before providing an overall conclusion is the safest approach to prevent misinterpretation or assumptions on behalf of the end-user.

Conclusion reporting

Although the standard which all forensic science disciplines should aim for is the reporting of conclusions in the form of likelihood ratios, this is only possible when the results of the analysis are quantitative, and when statistical databases are available from which to form the basis of a ratio. Areas such as DNA analysis and speaker comparison are well suited to this type of reporting as they meet both criteria. As it relates to audio authentication, although there are some specific analyses which are beginning to implement likelihood ratios, many are not suitable for reporting results in this manner.

The accepted method for reporting of authentication results, therefore, requires comparison against an accepted standard to be made. First, the standard baseline for comparison in relation to the authenticity of a recording is the most original version, as it is accepted that this is authentic. This is used as the point of comparison, because, if a recording is original, the possibility that editing of any form has occurred is eliminated. If this cannot be achieved, even if there are no traces of editing, the possibility will always exist that it may have been tampered with, and evidence of this went undetected due to the current limitations of the science.

Second, the term 'consistent with' is used to indicate the findings from the analysis stage. As the analyses do not result in any form of statistical measurement, none can be given when reporting the overall conclusion. Although some examiners may take it upon themselves to use arbitrary values when questioned in court, it is misleading, and even dangerous to provide any form of quantitative measurements such as a percentage or likelihood ratio to report findings, when the data does not exist with which to perform a statistical calculation. Only the following propositions should be considered:

P_1 – The exhibit is consistent with an original recording.
P_2 – The exhibit is inconsistent with an original recording.

With that in mind, there are three possible outcomes of an authentication examination (SWGDE, 2018):

1   Consistent with an original recording
2   Inconsistent with an original recording
3   Inconclusive

An examiner should not be scared to use the final proposition if the results are inconclusive. Where no information with regards to the device is available, the purported capture method, known editing techniques, and traces of known editing software must be considered. This shows the importance of a database of exemplar recordings, as it may be possible to provide a conclusion by finding an exhibit which is consistent with a recording within the database.

If there are artefacts present within a recording, the cause of which cannot be conclusively determined, these should be documented. There may be a number of potential causes, some of which are nefarious and others which are innocent. Where it is not possible to opine on the cause due to a lack of information or access to the purported device, a description of the artefact and the hypotheses for the cause of such should be provided. If it is felt that further information, such as exemplar recordings, access to the original recording device, or more information in relation to the original recording would allow further analysis which could potentially allow a reliable opinion to be given on the cause of the artefacts, this should also be stated.

It is also important to note that the conclusion is based on the analysis performed within the examination, as there are occasions when the analysis is limited due to factors such as the generation of the exhibit provided or information in relation to the original capture device. For example, if a recording was converted, but the original was no longer available, and no information was available in relation to the original device, it may still be possible to provide an opinion, but the factors on which the conclusion relies should be fully transparent. In the example case, it should be stated that analysis is limited and reasons for such given. These may include:

1   No information was known in relation to the purported capture device.
2   Access to the purported capture device was not available, and, as such, no exemplar recordings are available for comparison.
3   Access to the purported capture device was possible, but permission was not given for exemplar recordings to be created.
4   Access to the purported capture device was not available, and, as such, no exemplar recordings from the device were available for comparison, so exemplar recordings from the organisation's original media database were used.
5   The original recording is no longer available, and it is accepted that the recording is not a bit-stream copy of the original version.

The scale of conclusions should always be reported to aid in transparency and allow the reader to know the reason for the final conclusion.

Summary

Audio authentication is a field which is gaining significance in a world of ubiquitous editing software and new technologies which contain the potential to create convincing synthesised representations of the human voice. In order to perform authentication examinations, the examiner needs not only to understand acoustic and digital signal processing theory but also have a deep understanding of audio codecs and containers. Results are then reported categorically as to the consistency of an exhibit with an original recording, as it is accepted that by its very nature, an original recording cannot have been manipulated.

References

Allamanche, E., Geiger, R., Herre, J., and Sporer, T., 1999. MPEG-4 low delay audio coding based on the AAC codec. Paper presented at the 106th AES Convention, Audio Engineering Society, Munich, Germany.

American National Standards Institute, 2019. ATIS Telecom Glossary. Available at: glossary.atis.org

Apple, 2016. Introduction to QuickTime file format specification. Available at: developer.apple.com/library/archive/documentation/QuickTime/QTFF

Apple Inc., 2016. Appendix G: Audio priming: Handling encoder delay in AAC, Introduction to Quicktime File Format Specification. Available at: developer.apple.com/library/archive/documentation/QuickTime/QTFF/RevHistory/QT...

Archer, H., 2012. Quantifying effects of lossy compression on electric network frequency signals. Paper presented at the 46th International Conference: Audio Forensics, Audio Engineering Society, Denver, CO, USA.

Audio Engineering Society, 2012. AES recommended practice for forensic purposes: Managing recorded audio materials intended for examination. Available at: www.aes.org/publications/standards/search.cfm?docID=29

Axon, 2019. Axon camera video watermark timestamp. Available at: help.axon.com/hc/en-us/articles/115002746247-Axon-Camera-Video-Watermark-Timestamp

Berman, J., 2015. Analysis of zero-level sample padding of various MP3 codecs. National Center for Media Forensics, University of Colorado, Denver. CO.

Brixen, E.B., 2007a. Techniques for the authentication of digital audio recordings. Paper presented at the 122nd AES Convention, Audio Engineering Society, Vienna, Austria.

Brixen, E.B., 2007b. Further investigation into the ENF criterion for forensic authentication. Paper presented at the 123rd AES Convention, Audio Engineering Society, New York.

Cooper, A.J., 2006. Detection of copies of digital audio recordings for forensic purposes. PhD thesis. The Open University, Milton Keynes, UK.

Cooper, A.J., 2008. The electric network frequency (ENF) as an aid to authenticating forensic digital audio recordings: an automated approach. Paper presented at 33rd International Conference: Audio Forensics, Theory and Practice, Audio Engineering Society, Denver, CO, USA.

Cooper, A.J., 2010. Detecting butt-spliced edits in forensic digital audio recordings. Paper presented at 39th International Conference: Audio Forensics: Practices and Challenges, Audio Engineering Society, Hillerød, Denmark.

Cuccovillo, L., Aichroth, P., 2017. Increasing the temporal resolution of ENF analysis via harmonic distortion. Paper presented at the 2017 AES International Conference on Audio Forensics, Audio Engineering Society, Arlington, VA, USA.

Derrien, O., 2019. Detection of genuine lossless audio files: Application to the MPEG-AAC codec. Journal of the Audio Engineering Society 67, 116–123.

Donnelly, D. and Rust, B., 2005. The Fast Fourier Transform for experimentalists, part I: Concepts. Computer Science Engineering 7, 80–88.

Draghicescu, D., Pop, G., Burileanu, D., and Burileanu, C., 2015. GMM-based audio codec detection with application in forensics. Paper presented at the 2015 38th International Conference on Telecommunications and Signal Processing (TSP), IEEE, Prague, the Czech Republic.

ENFSI, 2005. Code of Conduct. Available at: enfsi.eu/wp-content/uploads/2016/09/code_of_conduct.pdf

Everest, F.A., 2006. Critical listening skills for audio professionals, 2nd ed. Artistpro, New York.

Fechner, N., Kirchner, M., 2014. The humming hum: Background noise as a carrier of ENF artifacts in mobile device audio recordings. Paper presented at the 2014 Eighth International Conference on IT Security Incident Management & IT Forensics (IMF), IEEE, Münster, Germany.

Fuller, D.B., 2013. Analysis of DC Offset in iOS devices for use in audio forensic examinations. National Center for Media Forensics University of Colorado, Denver, CO.

Galou, G. and Chollet, G., 2011. Synthetic voice forgery in the forensic context: a short tutorial. Available at: www.researchgate.net/publication/279833557_Synthetic_voice_forgery_in_the...

Google Deepmind, 2016. Wavenet: A generative model for raw audio. Available at: deepmind.com/blog/wavenet-generative-model-raw-audio

Grigoras, C., 2002. Digital audio recording analysis: The electric network criterion. Diamond Cut Productions, Hibernia, New Jersey, USA.

Grigoras, C., 2007. Applications of ENF analysis method in forensic authentication of digital audio and video recordings. Paper presented at 123 AES Convention, Audio Engineering Society, New York City, New York, USA.

Grigoras, C., 2010. Statistical tools for multimedia forensics. Paper presented at: 39th International Conference: Audio Forensics: Practices and Challenges, Audio Engineering Society, Hillerød, Denmark.

Grigoras, C. and Geradts, Z., 2019. Forensic multimedia authentication: Real-life challenges and solutions workshop. Available at: news.aafs.org/section-news/digital-multimedia-sciences-dms-november-2018

Grigoras, C., Rappaport, D., and Smith, J.M., 2012. Analytical framework for digital audio authentication. Paper presented at 46th International Conference: Audio Forensics, Audio Engineering Society, Denver, CO, USA.

Grigoras, C. and Smith, J.M., 2014. Quantization level analysis for forensic media authentication. Paper presented at the 54th International Conference: Audio Forensics, Audio Engineering Society, London.

Grigoras, C. and Smith, J.M., 2017. Large-scale test of digital audio file structure and format for forensic analysis. Paper presented at the AES Audio Forensics Conference, Audio Engineering Society, Arlington, VA.

Grigoras, C. and Smith, J.M., 2019. Forensic analysis of AAC encoding on Apple iPhone. Paper presented at the AES Audio Forensics Conference, Audio Engineering Society, Porto, Portugal.

Grigoras, C., Smith, J., and Jenkins, C., 2011. Advances in ENF database configuration for forensic authentication of digital media. Paper presented at 131st AES Convention, Audio Engineering Society, New York City, New York, USA.

Herre, J., 1999. Temporal noise shaping, quantization and coding methods in perceptual audio coding: a tutorial introduction. Paper presented at the 17th AES International Conference on High Quality Audio, Audio Engineering Society, Florence, Italy.

Hicsonmez, S., Sencar, H.T., and Avcibas, I., 2013. Audio codec identification from coded and transcoded audios. Digital Signal Processing 23, 1720–1730.

Huijbregtse, M. and Geradts, Z., 2009. Using the ENF criterion for determining the time of recording of short digital audio recordings. In Z.J.M.H. Geradts, K.Y., Franke, and C.J. Veenman (eds), Computational forensics. Springer, Berlin, pp. 116–124.

Izhaki, R., 2008. Mixing audio: Concepts, practices and tools. Elsevier Ltd, Oxford.

Kinnunen, T., Hautamaki, V., and Franti, P., 2006. On the use of long-term average spectrum in automatic speaker recognition. Paper presented at the International Symposium on Chinese Spoken Language Processing, Kent Ridge, Singapore.

Kitzing, P., 1986. LTAS criteria pertinent to the measurement of voice quality. Journal of Phonetics 14, 477–482.

Koenig, B.E., 1990. Authentication of forensic audio recordings. Journal of the Audio Engineering Society. 38, 3–33.

Koenig, B.E. and Lacey, D.S., 2012a. An inconclusive digital audio authenticity examination: A unique case. Journal of Forensic Sciences 57, 239–245.

Koenig, B.E. and Lacey, D.S., 2012b. Forensic authenticity analyses of the header data in re-encoded WMA files from small Olympus audio recorders. Journal of the Audio Engineering Society 60, 255–265.

Koenig, B.E., Lacey, D.S., Grigoras, C., Gali, S., and Smith, J.M., 2012. Evaluation of the average DC offset values for nine small digital audio recorders. Paper presented at the AES International Conference, Audio Engineering Society, Denver, CO, USA.

Kumar, R., Sotelo, J., Kumar, K., de Brébisson, A., and Bengio, Y., 2017. ObamaNet: Photo-realistic lip-sync from text. arXiv preprint arXiv:1801.01442.

Luo, D., Luo, W., Yang, R., and Huang, J., 2014. Identifying compression history of wave audio and its applications. ACM Transactions on Multimedia Computing, Communications and Applications 10.

Luo, D., Yang, R., and Huang, J., 2015. Identification of AMR decompressed audio. Digital Signal Processing 37, 85–91.

Microsoft Corporation and IBM Corporation, 1991. Multimedia programming interface and data specifications 1.0. Available at: www.tactilemedia.com/info/MCI_Control_Info.html

National Institute of Justice, Office of Justice Programs, 2004. Forensic examination of digital evidence: A guide for law enforcement: (378092004-001). U.S. Department of Justice, Washington, DC.

Nicolalde Rodriguez, D.P., Apolinario, J.A., and Biscainho, L.W.P., 2010. Audio authenticity: Detecting ENF discontinuity with high precision phase analysis. IEEE Transactions on Information Forensics and Security 5, 534–543.

Raissi, R., 2002. The theory behind MP3. Available at: www.mp3-tech.org/programmer/docs/index.php

Rappaport, D.L., 2012. Establishing a standard for digital audio authenticity: A critical analysis of tools, methodologies and challenges. National Center for Media Forensics, University of Colorado, Denver, CO.

Sanders, R.W., 2008. Digital audio authentication using the Electric Network Frequency. Paper presented at the 33rd International Conference: Audio Forensics, Audio Engineering Society, Denver, CO, USA.

Smith, J., Lacey, D., Koenig, B., and Grigoras, C., 2017. Triage approach for the forensic analysis of Apple iOS audio files recorded using the 'voice memos' app. Paper presented at 2017 AES International Conference on Audio Forensics, Audio Engineering Society.

Sotelo, J., Mehri, S., Kumar, K., Santos, J.F., Kastner, K., et al., 2017. CHAR2WAV: End-to-end speech synthesis. In Proceedings of ICLR Workshop, 2017, Toulon, France.

Sripada, P., 2006. MP3 decoder in theory and practice. Department of Signal Processing and Telecommunications, Blekinge Institute of Technology, Blekinge, Sweden.

Starkie, B., Findlow, G., Ho, K., Hui, A., Law, L., et al. 2002. Lyrebird™: Developing spoken dialog systems using examples, Telstra New Wave Pty Ltd, Victoria, Australia.

Strawn, C., 2009. Expanding the potential for GPS evidence acquisition. Small Scale Digital Device Forensics Journal 3.

SWGDE, 2016. Best practices for forensic audio. Available at: www.swgde.org/documents/published

SWGDE, 2018. Best practices for digital audio authentication 27. Available at: www.swgde.org/documents/published

Wang, Y. and Vilermo, M., 2002. The modified discrete cosine transform: Its implications for audio coding and error concealment. Paper presented at the AES 22nd International Conference on Virtual, Synthetic and Entertainment Audio, Audio Engineering Society, Helsinki, Finland.

Welch, P., 1967. The use of Fast Fourier Transform for the estimation of power spectra: A method based on time averaging over short, modified periodograms. IEEE Transactions on Audio Electroacoustics 15, 70–73.

Yang, R., Qu, Z., and Huang, J., 2012. Exposing MP3 audio forgeries using frame offsets. ACM Transactions on Multimedia Computing, Communications and Applications 8.

Zakariah, M., Khan, M.K., and Malik, H., 2018. Digital multimedia audio forensics: Past, present and future. Multimedia Tools Applications 77, 1009–1040.

Zjalic, J., Grigoras, C., and Smith, J., 2017. A low cost, cloud-based, portable, remote ENF system. Paper presented at 2017 AES International Conference on Audio Forensics, Audio Engineering Society, Arlington, Virginia, USA.

Chapter 12

Forensic reporting

Introduction

Once a forensic examination has been performed, a final report must be prepared based on the work conducted. The reporting aspect of forensics, which includes the methodology used and any conclusions rendered, can be considered to bridge the gap between a forensic laboratory and the courtroom. Reporting can be in the form of a written statement or through verbal testimony, but in the majority of legal systems, the latter cannot be performed without the former having been first introduced. In some scenarios, the evidence is accepted in a written format, with no obligation for the examiner to give verbal evidence. As there are numerous publications and training courses on the giving of oral evidence, this short chapter will focus on only the written form.

Purpose

The purpose of a written report is to present the methodology (and possibly the analysis and subsequent findings) of an examination to the end users. The report will then be considered by the instructing party and a decision made as to whether it supports their case and if they will be submitting it as evidence. A report needs to reach a balance between accessibility (thus allowing non-technical readers to understand the meaning of the findings and results of any analysis) while maintaining scientific integrity. Executing this balance is a skill in itself, as if too simplistic, it may be of no use as a non-expert would likely reach the same opinion. If too technical, the findings from the analysis may not be understood, potentially resulting in a back-and-forth question-and-answer session between the instructing party and the examiner to determine if the conclusions reached support or harm their case. In a worst-case scenario, the report may not be used, due to a lack of comprehension, or if the findings are misinterpreted by the reader. The inclusion of a glossary in which technical elements are described in laymen's terms is something which should be done to aid in the production of an accessible report. Another is to document all of the premises on which the conclusion is based to ensure transparency.

The quality of a report is the most visible evidence of an expert's professionalism, and as such, errors in punctuation, spelling, and grammar can suggest a carelessness which extends beyond the report writing and into their analysis. If an error has been made in the writing of the report, how can the end user be sure that no critical errors have been made during the analysis? Other factors which reduce the quality of a report include unnecessary repetition, an overly dense writing style, illogical structure, poor reasoning, and unsupported opinions (Appelbaum, 2010).

As reports vary based on the instruction, the topic, and the end user, there is no set method for writing a forensic report, but the principles of forensic science should always be applied. The report must also be independent of the party who instructed the work, and as such, any conclusions reached should be presented objectively and free of bias. For this to be ensured, all reports should be checked over by the author and peer-reviewed by at least one other examiner who is competent in the field to which the content of the report relates. A study of reports written by forensic psychologists (Grisso, 2010) documented the most common faults in the writing of expert reports as follows:

1 Opinions without sufficient explanations – stating opinions without explaining the basis of such.
2 Forensic purpose unclear – an inaccurate, inappropriate or missing purpose to the report.
3 Organisation problems – disorganised presentation of information.
4 Irrelevant data or opinions – inclusion of data which has no bearing on the purpose of the report.
5 Failure to consider alternative hypothesises – where data allows for alternative propositions, these were not considered when drawing conclusions.
6 Inadequate data – the opinion requires further data which was not obtained or requested.
7 Data and interpretation mixed – the mixing of facts and opinion, making it difficult to determine which is which.
8 Over-reliance on a single source of data – a conclusion which relies on a single source of data or single analysis.
9 Language problems – instances of biasing phrases or technical jargon.
10 Improper tests – use of inappropriate tests for the data type being analysed.

To combat the above, Witt (2010) performed research into the reasons the same errors appear time and time again in reports by different examiners, and the ways to mitigate against such. A checklist was proposed that prevents people from skipping steps (even when they think they remember them) and reduces the fallibility of human attention when performing mundane tasks, such as repetitive writing. The list consisted of checks concerning a report's organisation, the amount of technical jargon, and the connection between data and opinions.

Further advice is to set the report out in sections, so the report is clear, logical, and easy to follow (Allnutt and Chaplow, 2001). With the aforementioned points in mind, the following provides an example structure for a forensic report, but it is essential that other error mitigation processes, such as peer review and report checklists are implemented to ensure reports of the highest quality are consistently produced.

Structure

In terms of structure, combining the tried and tested IMRAD format (Introduction, Methods, Results, and Discussion) used for the publication of scientific research with a sequence similar to that of courtroom testimony aids the examiner, should they be requested to give verbal evidence. It also ensures the report is presented in a logical and scientific manner to the end user. There will be occasions when some sections are omitted, and others added, based on the work performed, and different legal systems will also have elements which must be included in the report for it to be accepted as evidence. For example, in the UK, there are different statements of declaration depending on whether the evidence is being submitted in a criminal or civil case. It is the role of the examiner to decide which sections are appropriate based on the legal system and the type of work performed.

Cover page

The cover page should include information, such as the case reference, type of work performed, name of the instructing party, the date the analysis took place, address of the laboratory at which the analysis took place, the date the report was issued, and the name of the examiner. This allows the reader to obtain information regarding the contents of the report quickly and easily.

Contents

As with all documents which are over a few pages in length, a contents page is required for ease of navigation.

Introduction

Examiner summary

A brief, one-paragraph summary of the examiner's qualifications and experience to provide the reader with confidence that the examiner is an expert in the topic on which the report is based.

Summary of those involved

A summary of those involved and their role in the work conducted. Generally, this will be the examiner and the individual who peer-reviewed the work, but may include others involved in the preparation or analysis stages.

Information regarding instruction

A summary of the instruction received by the instructing party, taken verbatim from the letter of instruction where possible to aid in transparency and prevent any misinterpretation.

Points of instruction

The information provided by the instructing party refined and focused as a distinct set of instructions. The foundation of these points is essential as they are the basis for the direction of the analysis. If these are incorrect, unclear, or omitted, there is a potential for addressing irrelevant issues and not addressing the reasons the examination was instructed.

Purpose of the report

The purpose of the analysis performed, for example, to assist the court, or to assist an internal investigation.

Conflicts of interest

Any conflicts of interest should be documented so considerations can be made by the end users as to whether these may cause bias and potentially influence any conclusions reached. These may relate to associations with parties involved in the case and the tools used.

Scope

Detailing both what is within the scope of the report and what is determined to be outside of scope.

Facts

A list of all the facts which may influence the examination and the results, including those provided by the instructing party upon instruction, which are pertinent to the analysis. Documenting them is essential in ensuring there is a clear distinction between fact and opinion, so there is no misinterpretation by the reader. An example may be that the exhibit provided is not a bit-stream copy and the reasons why the examination continued without a copy of the original evidence.

Assumptions

A list of any assumptions made within the analysis and subsequent report which may influence the examination and results, and the reason for these assumptions. Continuing the example of the non-bit-stream copy, an assumption could be that the exhibit may be of lower quality in comparison to the original version as it is not a bit-stream copy of such.

Limitations

A list of the limitations of the analysis and how these can affect findings or conclusions. Further continuing the example from the previous sections, one limitation would be that analysis performed by another examiner who had access to a bit-stream copy might result in different conclusions due to quality differences.

Preparatory analysis

Any analysis which was performed in preparation for the principal analysis stage.

Methodology

Statement of methodology

The method used for the analysis should be documented to allow the reader to understand the approach to the work performed and for another examiner of the same level of competence to obtain the same results if reproducing the analysis.

Results

Analysis and results

The body of the analysis should be documented to provide the instructing party with all results and the meaning of such in laymen's terms.

Discussion

Opinion

When the rendering of a conclusion is required, any opinions and the premises from the analysis stage on which they are based should be documented. Premises are facts and findings that make a point. When deductive reasoning is used (in which conclusions pertain only to the premises and no assumptions based on such are made), it stands to reason that if the premises are reliable, so too must be the conclusions (United Nations Office on Drugs and Crime, 2011). An example of deductive reasoning pertaining to the rendering of an opinion in an audio authentication is as follows:

Conclusion:

- The exhibit is inconsistent with an original recording.

Premises:

- The recording is in a WAV PCM format.
- It is purported to have been recorded with device 'A.'
- Device 'A' does not have the capabilities to record audio using the WAV PCM format.

The above is a crude example, as in a real report, the conclusion would be supported by a far more extensive number of premises. The premises could also be condensed into shorter, concise statements to aid in more succinct reporting.

Statement

A standardised statement for use in forensic reports is usually published in accordance with the legal system in which the work is being performed. Analogous to the small print in a document, it covers areas such as the obligations that the expert is in agreement with, a statement to the effect that all of the information within the document is true to the best of their knowledge, and the understanding of their duty to the court rather than the party who instructed them (Forensic Science Regulator, 2019).

Appendix

Glossary

As audio forensics is a technical discipline, it is necessary to clearly define any uncommon terms used within the report in a manner that can easily be understood by laymen.

Tools

All tools used, whether software or hardware, their versions, and their application in the work performed should be documented to allow another examiner of the same level of competence to reproduce the analysis.

Conclusion reporting scale

Where there is a range of opinions among experts, both the scale for reporting results pertinent to the analysis and the guidance on the rendering of a conclusion using said scale should be included to aid in transparency.

Exhibits reviewed

The exhibits reviewed during the examination should be documented, so another examiner is able to reproduce the analysis, and also to ensure the chain of custody is maintained. If there were exhibits which were requested but were not provided, a list of those which would have aided the examination should be included to support the previously stated limitations.

Exhibits generated

A list of the filenames and checksums of any exhibits generated and provided to the instructing party should be documented to start a chain of custody. This ensures any changes made to the generated exhibits after they have been delivered to the instructing party can be detected.

Bibliography

A list of all materials which were consulted and relied upon for the work performed should be included to support findings and allow the reader to obtain further information should they so wish.

Examiner qualifications and experience

An extended summary of the qualifications and experience in the pertinent subject area, which allows the expert to give an opinion, must be provided, again to support the reliability of the work and any conclusions reached.

Summary

The reporting of results is the final step of any examination and is just as important as the earlier stages. Effective communication is vital to prevent misinterpretations and to ensure all of the findings are presented to the trier of facts to assist them in arriving at a just, fair, and considered verdict. Doing great work is of little consequence if it is poorly communicated, as the report may be the only representation of the work performed that the end user will access. Following a scientific format combined with the logical structure, which the verbal courtroom testimony is likely to follow, is a practical approach to ensure all the information is not only provided in a scientific manner but is also accessible to the reader.

References

Allnutt, S.H. and Chaplow, D., 2001. General principles of forensic report writing. The Australian and New Zealand Journal of Psychiatry 34, 980–987.

Appelbaum, K.L., 2010. Commentary: The art of forensic report writing. The Journal of the American Academy of Psychiatry and the Law 38, 32–42.

Forensic Science Regulator, 2019. Expert report content (No. Issue 3). Available at: www.gov.uk/government/publications/expert-report-content-issue-3

Grisso, T., 2010. Guidance for improving forensic reports: A review of common errors. Open Access Journal of Forensic Psychology 2.

United Nations Office on Drugs and Crime, 2011. Criminal intelligence: Manual for managers. United Nations, New York.

Witt, P.H., 2010. Forensic report checklist. Open Access Journal of Forensic Psychology 2, 233–240.

Chapter 13

Research

Introduction

Research plays both a necessary and critical role in all scientific endeavours, from its beginnings in ancient Greece and Rome, where deductive logic (the drawing of conclusions based only on a set of premises) was first refined as a science, through to modern-day forensic laboratories (Franceschetti, 2017). Without the research performed by individual scientists and organisations into audio forensics, no such field would exist. It is the research conducted by Bell Labs to develop the spectrogram during the Second World War that paved the way for the field, and it is essential that research continues and further methods are developed as technology continues to evolve and new devices, software, and audio formats enter the market. Audio forensics must not only stay relevant to ensure that previously developed methods are adapted to the latest technology, but it must also create new approaches to ensure the maximum amount of information is extracted from every recording. In developing techniques, the arsenal of methods which are already available is increased, thus improving the reliability of results and providing a more substantial degree of confidence in any conclusions drawn.

Although there are a few commercial organisations which do perform research, such as those that develop software and processes for audio enhancement, the majority of research is performed within academia and published in the form of posters, conference papers, and articles in peer-reviewed journals (Tondello, 2018). The reasons for the limited amount of research by commercial entities are many, including a lack of funds through to backlogs in casework, all of which severely limit the resources and possibilities of research being undertaken.

Research may be highly specific with no application to casework, or extremely broad but can be applied to casework, or a mixture. By its nature, a method that is published in a journal is peer-reviewed and thus can be considered reliable, providing the results are made available, although all processes should be validated and repeatedly verified in each laboratory to show that they are fit for purpose (Grigoras and Geradts, 2019). Every research study published adds another piece to the puzzle, so even if it has no application to casework, the next researcher may take the findings and apply them to their own research which might be applicable to casework. It may take several iterations until a method is finally proposed which is appropriate to casework.

It is hoped that the inclusion of this chapter will help the reader gain an understanding of research and how to conduct it scientifically, resulting in further advances in the field.

The scientific method

The scientific method is the logical approach to solving a problem as objectively as possible by moving through a series of steps, often using a process of hypothesis testing (Triplett and Cooney, 2006). Although there are many variations of the method, a good breakdown of concise steps is presented by Reznicek et al. (2010) as follows:

1   Make an observation.
2   State the problem or question.
3   Generate a hypothesis.
4   Conduct experiments and collect data.
5   Generate conclusions based on the data.
6   Confirm the process and conclusions through replication.
7   Record or present the conclusions.

Observation

The basis for all research is an observation, which often relates to a problem that needs to be solved. In the case of the spectrogram research, this was the need to determine the difference between allied and enemy submarines during the Second World War, based on the differences in frequencies emitted by the two types. In the case of quantisation level analysis, it was that there was no way of determining if a recording had been transcoded from one of a lower bit depth, which led to observations relating to differences in the quantisation level distribution when recordings are converted from lower to higher bit depths.

Literature review

Once an observation has been made, a literature review must take place to discover the conclusions reached by other researchers with regards to the pertinent topic, and any other research or methods that may apply to the study. It can also prevent the repetition of research that has already been performed, potentially saving both money and time.

Hypothesis

The next stage is the development of a hypothesis. The purpose of a hypothesis is to provide direction for the testing stage that follows. It is essentially an educated guess based on existing data (which may have been collected by the researcher during some preliminary experiments) and experimental results from other published research (acquired from the literature review stage) (Franceschetti, 2017). The hypothesis takes the form of a statement that predicts the relationship between two or more variables with regards to an observation, of which there are three types, dependent on the stage at which the research has reached (ibid.).

Hypothesis: An educated guess.

Working hypothesis: A hypothesis which is accepted for the time being based on experiments relating to the initial hypothesis.

Scientific theory: A theory which has been tested and confirmed repeatedly.

A hypothesis can also exist in the form of a proposition, such as a conditional statement like 'if p, then q.' A null hypothesis is every other possible outcome other than that stated within the hypothesis. It essentially indicates that there is either no relationship between the variables or that the relationship is not as expected. The variables on which the prediction is based can be either dependent or independent. An independent variable is an element which is manipulated and thus causes the outcome, while the dependent variable is the element which is affected by the independent variable.

The following is an example of a simple observation and the associated variables and hypothesis:

Observation: If I flip a coin, it always seems to land on heads.

Independent variable: The flipping of a coin.

Dependent variable: The result being heads.

Hypothesis: A coin will land on heads 100 per cent of the time when flipped.

Null hypothesis: A coin will land on heads less than 100 per cent of the time when flipped.

Experiment

Once a hypothesis has been defined, its validity must be tested through experimentation. This usually involves repeated testing to create a body of evidence which can support, refute, or refine the hypothesis.

To perform an experiment, a set of data is required, of which there are generally two types. The first is that which has been created manually by the researcher or somebody within the research group. This can be extremely time-consuming as original recordings would first have to be captured before the manipulation of a variable is performed manually by a user. The second is automated, which still requires the capture of original recordings, but the manipulation can then be performed through the use of algorithms. This method is ideal when a large amount of data is required, such as in machine learning tasks, and these datasets are often shared between researchers. Various datasets exist for audio forensics, such as that created by Vryzas et al. (2018), in which recordings in a WAV format were edited automatically, converted to a different format (MP3, AAC, FLAC, and AMR) and converted back to a WAV format. A dataset like this is suitable for the testing of both codec and local edit detection methods.

For experiments and thus the results to be reliable, there are several requirements for a dataset:

1   Applicable
    Although obvious, this is the first requirement as every stage beyond this would be redundant without first ensuring the data is relevant to the testing being performed. For example, if the hypothesis relates to the number of apples falling from a tree, but there are no apple trees so pears are measured, the dataset cannot be considered to apply to the hypothesis.

2   Variables
Aside from being independent and dependent, there are two other classifications for variables used within scientific experiments, those of controlled and uncontrolled.

If the data being used has various uncontrolled variables, it would likely be impossible to reliably determine the cause of any observations. All variables should, therefore, be controlled, and only the independent variable manipulated. For example, if the hypothesis relates to differences in the number of audible clicks based on a recording device, variables such as the recording length, format, and type of sound being captured should all be controlled, while the recording device variable is manipulated. In doing so, it can be ensured any findings are a result of the recording device and not the length, format or character of the sound being captured.

3   Groups
Two types of data should be created, namely, experimental and control.

In the experimental data group, the independent variable is manipulated, so its effects on the dependent variable can be measured. In the control data group, the independent variable remains unchanged, and so it can be determined if the effects seen in the dependent variable are due to the independent variable or have another cause. The control group essentially provides a baseline to be measured against.

4   Large enough to allow for rigorous testing
Without testing many pieces of data, it is difficult for the two scientific principles of repeatability and reproducibility discussed in earlier chapters to be met. There will, therefore, be a limited level of confidence in the results obtained from a small data set as the hypothesis has not been given the opportunity to fail, the limits of the hypothesis are not determined, and factors such as the standard deviation and error rate cannot be accurately calculated. Although in general, the more data tested, the more reliable the conclusions drawn, there is a point where the benefits of more data become negligible, or for practical reasons cannot be obtained. The amount of data required is different for each hypothesis and so must be considered on a case-by-case basis.

5   Reliable
Results are only as reliable as the data tested, and as such, all of the previously documented factors apply, but other elements must also be considered. It may be time-saving and convenient to obtain data from other sources (such as other examiners, friends, or the internet), but reliability may be comprised in doing so. Can you be confident in the results of research if the capture method and authenticity cannot be verified? It is always best practice to create the dataset from scratch or obtain one from a trusted and reliable source, such as previous research projects where the collection method is documented and the data has been made available for public use.

Once the dataset collection method has been defined, a protocol for collection must be created to ensure that all of the above are met and that there is transparency should anybody else want to review the collection method or use the dataset. This is essentially a document consisting of a list of the data which will be collected and a standard operating procedure to ensure reproducibility.

Results

Once the data has been collected, it can then be analysed. The type of analysis will be highly dependent on both the hypothesis and the data, but it should be applicable to the hypothesis. The results of such are then documented and presented objectively.

Discussion

Finally, a decision can be made as to whether the hypothesis is correct, based on the results. As the null hypothesis is the inverse of the hypothesis, providing the previous stages have been performed correctly, an inconclusive result should rarely occur. Opinion can also be given where necessary on specific findings and their reasons, and suggestions for further research made. The hypothesis can then be refined, and further experiments designed. Scientific research is an ongoing cycle, in which theories are only accepted until additional data is acquired, at which point the theory is re-evaluated.

Reporting

The reporting of the research will depend on its intended use. For example, organisations who have performed a study to improve their own products will most likely restrict the reporting of results internally within their organisation to gain an advantage over their competitors (Polidoro and Theeke, 2012). Another route is for the results to be published in a journal, and it is the responsibly of the scientist to determine the most suitable journal for their research. Journal papers follow a standardised structure defined as IMRAD, namely, Introduction, Methodology, Results and Discussion. IMRAD's development began after the Second World War when the number of advances made by medical research laboratories increased, and journal editors began to demand concise (to keep printing costs down) and well-written (as there was now a lot of competition) manuscripts. IMRAD was seen by some journals as the most logical way to communicate scientific results, and others soon followed suit (Day and Gastel, 2012).

Once submitted for publication, the research undergoes peer-review by a minimum of two of the scientist's peers. These individuals are selected by the editor of the journal as being qualified to opine on the topic to which the research pertains. Peer review is essential in identifying mistakes, whether justifiable and unknown to the author, or due to careless errors during the research. They are also vital in mitigating against fraud, where results are doctored to overstate the findings of a piece of research (Crawford and Stucki, 1990). Integrity is critical for science as a whole, and it could be argued even more so for science applied to forensics, as, in the most extreme case, people have been wrongly imprisoned for a number of years before being exonerated when the techniques on which results were based were found to be unreliable. For the same reasons, it is crucial to measure errors in forensic science methods as the probative value of evidence is strongly linked to the error rate. Without this information, triers of fact have no scientifically meaningful way to assign a weight of reliability to each piece of evidence. Where possible, they should also be provided with the rate of false positives for the methodology in question, to prevent presumption, the discredit of forensic science portrayed in the media, and unscientific claims made by supposed experts in the courtroom (Koehler, 2013).

Regardless of whether the hypothesis was found to be correct or not, if the research is performed scientifically, there are no wrong conclusions. There is a tendency in science to only report positive results, but by publishing research in which the hypothesis was found to be incorrect, it allows others to review methods which failed, and possibly improve on it or save time by not repeating the same process.

Summary

Without the research performed by our predecessors and peers, the forensic audio field would likely not exist. The Red Queen's opinion of the race (Carroll, 1871) in *Through the Looking Glass,* goes to the heart of forensic science:

> 'Well, in our country,' said Alice, still panting a little, 'you'd generally get to somewhere else – if you run very fast for a long time, as we've been doing.'
>
> 'A slow sort of country!' said the Queen. 'Now, here, you see, it takes all the running you can do, to keep in the same place.'

As new technologies emerge and new anti-forensics techniques are created, research is essential for us to just stay in the same place. It requires forethought to predict the areas which may create problems in our field in the future, and the ability to react to new technologies and techniques. It also provides us with logical ways of thinking and a greater understanding of methods and their limitations. which can serve to reduce bias, errors, and misinterpretations of forensic casework.

References

Carroll, L., 1871. Through the looking glass. Macmillan, London.

Crawford, S. and Stucki, L., 1990. Peer review and the changing research record. Journal of the American Society for Information Science 41, 223–228.

Day, R. and Gastel, B., 2012. How to write and publish a scientific paper, 7th edn. Cambridge University Press, Cambridge.

Franceschetti, D.R. (ed.), 2017. Principles of scientific research. Salem Press, Ipswich, MA.

Grigoras, C. and Geradts, Z., 2019. Forensic multimedia authentication: Real-life challenges and solutions workshop. Available at: news.aafs.org/section-news/digital-multimedia-sciences-dms-november-2018

Koehler, J.J., 2013. Proficiency tests to estimate error rates in the forensic sciences. Law, Probability and Risk 12, 89–98.

Polidoro, F. and Theeke, M., 2012. Getting competition down to a science: The effects of technological competition on firms' scientific publications. Organization Science 23, 1135–1153.

Reznicek, M., Ruth, R.M., and Schilens, D.M., 2010. ACE-V and the scientific method. Journal of Forensic Identification 60, 87–103.

Tondello, G.F., 2018. How to publish research results for academic and non-academic audiences. XRDS: Crossroads, The ACM Magazine for Students. Available at: blog.xrds.acm.org/2018/02/how-to-publish-about-your-research-results-for...

Triplett, M. and Cooney, L., 2006. The etiology of ACE-V and its proper use: An exploration of the relationship between ACE-V and the scientific method of hypothesis testing. Journal of Forensic Identification 56.

Vryzas, N., Katsaounidou, A., Kotsakis, R., Dimoulas, C., and Kalliris, G., 2018. Investigation of audio tampering in broadcast content. Paper presented at the 144th AES Convention, Audio Engineering Society, Milan, Italy.

Glossary

a-law: Companding technique to optimise the dynamic range of a system by reducing it from 13 bits to 8 bits of logarithmic PCM data. Used in Europe for telephony communication.

Accreditation: Process of verifying that a laboratory has an appropriate quality management system and can perform methods within a specific scope to an international standard.

Acoustics: Science of sound, including the production, transmission, and effects.

Adaptive Differential Pulse Code Modulation (ADPCM): Variation of PCM that encodes the difference between samples rather than the sample values. The quantisation step size adapts to changes in amplitude.

Adaptive MultiRate (AMR): Lossy compression codec optimised for the encoding of speech signals.

Admissible: Accepted as evidence within the legal system.

Aliasing: When an analogue signal is sampled, frequencies which exceed half the sampling rate will appear within the lower half of the signal, thus being 'aliases' of higher frequency signals if a low pass filter is not applied.

Alternating Current (AC): Flow of electrons sourced from mains socket outlets which change direction periodically.

Amplitude: Signal level of a waveform.

Analogue audio: Audio recordings which replicate the continuous nature of original sound waves, as opposed to the discrete nature of digital audio.

Analogue to Digital Converter (ADC): System for converting an analogue signal to the digital domain.

Anti-alias filter: Low pass filter implemented before the analogue to digital conversion to prevent signal components which exceed half the sample rate from entering the converter, which would result in aliasing.

Anti-forensics: Application of a process to inhibit or prevent the effectiveness of a forensic science examination.

Artefact: Feature of an audio recording resulting from human activity or a technical process.

ASCII: Acronym for American Standard Code for Information Interchange. An international standard language used for metadata within file containers.

Attenuation: Reduction of a signal's amplitude in the time or frequency domain.

Audio forensics: Application of the science of audio to legal matters.

Audio Interchange File Format (AIFF): Uncompressed audio format developed by the Apple Corporation.

Auditory fatigue: Reduction in the sensitivity of the human auditory system caused by prolonged exposure to sound.

Authentication: Process of substantiating that an exhibit is a true representation of what it purports to be.

Bandpass filter: Filter that attenuates signals either side of a band defined by a centre frequency, allowing only frequencies within that band to pass.

Bandwidth: Range of frequencies contained within an audio signal.

Basilar membrane: Small organ in the ear that acts as a spectrum analyser through different fibres in the cochlear nerve being stimulated by different frequencies.

Bel: Logarithmic measurement coined by Alexander Graham Bell to express the ratio of one quantity against another.

Binary: System which exists in only two states, zero or one.

Bit: Measurement unit of binary data, which can be either zero or one.

Bit depth: Number of bits available per sample to represent the amplitude information when using Pulse Code Modulation.

Bit rate: Number of bits transferred or replayed per period, commonly denoted in bits per second. Due to the large numbers which this measurement can reach, it is usually presented in Kilobits or Megabits.

Bit-stream copy: Complete and exact digital copy of all bits of data.

Butt splice: Meeting of two samples from the ends (or butts) of two regions of consecutive samples.

Buzzing: Additive periodic sinusoid composed of a large number of harmonics.

Byte: Collective term for eight bits of digital data.

Capture: Process of recording a sound.

Capture device: Device used for the capture of sound.

Cepstrum: Inverse Fourier Transform of the logarithm of the Fourier Transform.

Chain of custody: Chronological documentation of the movement, location, and possession of an exhibit.

Channel: Path carrying audio data. Most commonly used in conjunction with mono and stereo.

Clipping: Signal level which exceeds the capabilities of the system, resulting in the level represented by the maximum value available within the system.

Codec: Program capable of encoding and decoding of digital data. Codecs encode a stream or signal for transmission, storage, or encryption and decode it for listening.

Cognitive bias:: Human inclination towards a preconceived or unsupported outcome.

Companding: Portmanteau of compressing and expanding used to describe the process of compressing a signal before returning it (expanding) to a format similar to the pre-compressed version.

Compression (acoustic): Increase in pressure within a medium. The opposite of rarefaction.

Compression (digital): Processing of data to reduce the amount of data required for the storage of such.

Compressor: Hardware device or computer software used to reduce the dynamic range of a signal by attenuating the signal once it exceeds a threshold by a specific ratio. The attack and release control the speed at which this occurs.

Container: File structure used to store multimedia data streams and information to ensure playback can be achieved. There are various standardised container structures available.

Critical bands: Series of bandwidths which stimulate separate regions of the basilar membrane. Tones which exist within a single critical band are not clearly identifiable.

Cyclic Redundancy Check (CRC): Digital data error detection technique in which checksums are calculated for blocks of data during encoding and stored in the file. Upon decoding, the checksums are again calculated and compared against those stored in the file.

Data: Information that can be transmitted for processing.

DC offset/bias: Mean amplitude of a signal. Commonly measured in quantisation levels or as a percentage.

Decibel (dB): Logarithmic measurement used to define the ratio between the amplitude of two signals.

Decoder: Algorithm to convert digital data stored in an audio file into audio information.

Deletion: Removal of a sample or samples from a larger sequence of such.

Delta modulation: Single bit audio encoding technique performed at high sample rates in which the value of each bit is determined only by the direction (increase or decrease) of the current sample in relation to the previous sample value.

Digital audio: Sound stored as a series of discrete samples.

Digital signal processing (DSP): Generic term to describe the processing of a digital signal.

Digital to Analogue Converter (DAC): System for converting a digital representation of a signal to the analogue domain.

Digital versatile disc (DVD): Optical format used for the storage of digital data.

Direct current (DC): Flow of electrons in a single direction produced by batteries.

Direct signal: Component of a signal which is not reflected or absorbed before reaching the receiver.

Discrete Fourier Transform (DFT): Fourier Transform dealing with time-domain signals that are both discrete and periodic.

Distortion: Undesired components which are associated with the original signal.

Dither: Addition of noise to a signal to randomise the quantisation error, thus reducing the perception of quantisation noise.

Dynamic range: Difference between the largest and smallest amplitudes within a signal.

Early reflections: Initial signal reflections following the propagation of an acoustic sound in a reflective environment.

Encoder: Algorithm to convert audio into digital data for storage.

Energetic masking: Masking which occurs within the acoustic environment due to the masker physically interacting with the maskee.

Enhancement: Process of improving the perceptual quality of a signal.

Entropy encoding: Type of lossless compression which encodes frequently occurring sequences with a lower number of bits, and rarely occurring sequences with larger numbers of bits.

Equalisation: Process of adjusting the frequency composition of a signal through attenuation and boosting of frequencies.

Equaliser: Hardware or software tool used to perform equalisation.

Equal loudness contours: Term used for a plot visualising frequency and loudness perception by the human auditory system, the first of which was called the Fletcher Munson curve.

Exhibit: Item of evidence. In the digital audio forensics field, this is generally an audio recording.

Extraction: Process of extracting data from a source.

Extrapolation: Creation of new samples outside the range of a signals original samples.

Fast Fourier Transform (FFT): Highly efficient algorithm used to calculate the Discrete Fourier Transform.

File footer: Final section within a digital multimedia file.

File header: First section within a digital multimedia file. It contains identifying information and potential metadata about the contents of the file.

Filename: Alphanumeric name used to identify a file within a file system.

File rewrapping: Process of transferring raw multimedia data to a new container.

Filter: Algorithm which separates signal components based on frequency by allowing some frequencies to pass and others to be attenuated.

Filter bank: Series of bandpass filters that separates the input signal into several signals.

Forensic report: Document detailing the methodology, results and conclusions of a forensic examination.

Forensic science: Application of science to legal matters.

Format: Used to describe the codec, or more commonly, the container, of a digital audio file.

Format conversion: Transfer information or data from one format to another.

Fourier series: Method for representing a periodic function (such as an acoustic waveform) as a series of sine and cosine waves.

Frames: Result of dividing a digital audio signal into regions consisting of equal numbers of samples.

Frequency: Number of cycles per second, measured in Hertz.

Frequency domain: Representation of a signal where the frequency is the independent variable.

Frequency resolution: Measurement pertaining to the spectral domain, determined by the width of the main lobe. The narrower the width of the lobe, the higher the resolution.

Fundamental frequency: Lowest frequency component of a periodic waveform.

Gain: Degree of amplification or attenuation of a signal, usually measured in dB.

Guideline: Recommended practice that allows some discretion as to its interpretation and implementation.

Hard Disk Drive (HDD): Physical device consisting of rotating platters used to store digital data.

Harmonic: Frequency which is a positive integer multiple of the fundamental frequency.

Hash checksum: Numerical value calculated through a mathematical function to verify the integrity of a digital file through comparison against other checksums.

Headphones: Device worn over the head or in the ears to playback sound to the user.

Hertz: Measurement of the number of samples per second.

Hex editor: Software program used to view and edit the hexadecimal representation of a digital file.

Hexadecimal: Base 16 system used to represent numbers.

High pass filter: Filter which attenuates the amplitude of frequencies below a defined cut-off frequency while letting those above it pass unaffected.

Histogram: Plot of frequency distribution to visualise the number of occurrences of an event.

Huffman encoding: Form of entropy encoding based on the frequency of occurrences of information within a signal.

Hum: Additive periodic sinusoid signal with a range limited to the lower frequencies.

Human Auditory System (HAS): Biological system used for sound perception by humans.

Hypothesis: Prediction of the relationship between two or more variables.

ID3: System for adding metadata to files, most commonly used within the MP3 format.

Inadmissible: Not accepted as evidence by a legal system.

Individuality principle: Principle that every object in the universe is unique and on which all forensic identification tasks are based.

Informational masking: Masking of a signal which occurs within the receiver rather than within the acoustic environment.

Intelligibility: Measure of how well speech can be understood, commonly measured as a rate of correctly transcribed words.

Interaural Intensity Difference (IID): Difference in the magnitude of an acoustic signal at two or more receivers.

Interaural Time Difference (ITD): Difference in the time taken for an acoustic signal to reach two or more receivers.

Inter Laboratory Comparison (ILC): Process of comparing the results of a method achieved by separate laboratories.

International Organization for Standardization (ISO): Standards organisation comprised of various international representatives.

Interpolation: Creation of new samples inside the range of a signal's original samples.

ISO 17025: Standard which lays out the competency requirements for test and calibration laboratories.

Journal: Scientific publication containing peer-reviewed research articles.

Late reflections: Succession of echoes with a diminishing intensity which arrive after the direct and early reflection components of an acoustic signal.

Least Significant Bit (LSB): Bit in a binary number which has the least effect on its value.

Limiter: Hardware or software which limits the amplitude of a signal above a defined threshold.

Linear Pulse Code Modulation (LPCM): Form of lossless encoding in which the quantisation levels are linearly distributed, and the values of such are proportional to the amplitude.

Long-Term Average Spectrum (LTAS): Representation of a signal calculated from the average power spectral density of a series of overlapped FFTs.

Long-Term Average Sorted Spectrum (LTASS): Representation of the results of a Long-Term Average Spectrum calculation sorted in descending order.

Lossless compression: Method of data reduction in which no information is lost and can be retrieved in its original format.

Lossy compression: Method of data reduction in which information is removed and cannot be retrieved.

Loudness: Perceived volume of an audio signal.

Low pass filter: Filter which attenuates the amplitude of frequencies above a defined cut-off frequency while letting those below it pass unaffected.

Maskee: Signal component which becomes less audible during masking.

Masker: Signal component which causes another (the maskee) to become less audible during masking.

Masking: Process of one component of a signal causing another to become less audible.

Metadata: Data that provides information about another piece of data.

Mean Opinion Score (MOS): Methodology for reporting the quality or intelligibility of an audio recording through the averaging of scores provided by a panel of participants in relation to various features.

Microphone: Transducer used to convert acoustic energy to an electrical signal.

Modified Discrete Cosine Transform (MDCT): Lapped transform based on the type-IV discrete cosine transform where the last half of one block coincides with the first half of the next. Used by a number of audio codecs including MP3, AAC, AC:3, Vorbis, and WMA.

Monoaural: Audio that exists in a single channel or stream.

Most Significant Bit (MSB): Bit in a binary number which has the most effect on its value.

Motion Picture Experts Group (MPEG): Subgroup of ISO focused on multimedia.

MP3: Lossy compression technique for digital audio, using perceptual encoding to achieve reduced file sizes.

Multimedia: Collective term for content that includes audio and imagery.

Native file format: Original format of a file.

Noise: Part of the signal which is not desired.

Non-simultaneous masking: Masking which occurs when the masker component exists before or after the maskee component in the time domain.

Normalisation: Transposing of all quantisation levels within a signal by the same amount of gain.

Nyquist frequency: Maximum frequency component that can be accurately represented within a discrete signal without any artefacts, defined as half the sampling rate.

Nyquist Theorem: Theory that states a system must have a sample rate at least twice the highest frequency within the desired signal for it to be accurately represented by a discrete signal.

Original: Recording made simultaneously with the acoustic events it purports to have recorded, made in a manner fully and completely consistent with the methods of recording claimed by the party who produced the recording and free from unexplained artefacts, alterations, additions, deletions, or edits.

Overlapping: Process of overlapping frames of audio data (either in the time or frequency domain) to ensure information which bridges two frames is not lost.

Passband: Frequency components which pass unaffected from the input to the output of a filter.

Payload: Section within the structure of an audio file which pertains to the audio data.

Peer review: Evaluation of a research paper or report by an independent party to evaluate methods and conclusions prior to publication.

Perceptual encoding: Coding of audio using models based on the human auditory system to increase efficiency and reduce redundancy.

Permanent Threshold Shift (PTS): Condition caused by over-exposure to sound which results in a permanent upwards shift of the hearing threshold and subsequently reduced sensitivity of the human auditory system.

Phase: Relationships between individual frequency components of a signal.

Post-masking: Masking which occurs when the masker signal occurs after the maskee signal in the time domain.

Power: Amount of energy transferred in a specific period, generally measured in Watts.

Power Spectral Density (PSD): Measurement of the energy of a signal as a function of frequency.

Pre-masking: Masking which occurs when the masker signal occurs before the maskee signal in the time domain.

Proficiency testing: Testing of competency in a specific task by comparing results against known outcomes.

Proprietary file format: File format unique to a specific manufacturer or product.

Pulse Code Modulation (PCM): Technique used to represent an infinite analogue signal discretely through the processes of sampling and quantisation.

Quality: Preference of a listener for one sound over another.

Quality assurance: Systematic activities within a quality system implemented to provide confidence that an entity will meet defined quality standards.

Quality factor (Q): Parameter of a filter which defines its bandwidth. The higher the Q parameter, the narrower the range of frequencies affected by a process.

Quantisation: Process of mapping a large number of continuous values to a smaller, discrete set.

Quantisation error: Error introduced during the quantisation process.

Quantisation noise: Noise introduced into a discrete signal by quantisation error.

Rarefaction: Decrease in pressure within a medium. The opposite of compression.

Reflection: Sound that has bounced off a surface within the acoustic capture environment.

Reproducibility: Degree to which a process yields the same results during repeated trials.

Reverberation: Collection of reflections caused by an audio signal bouncing off surfaces within the acoustic capture environment.

Root Mean Square (RMS): Method for calculating the mean amplitude of a period by squaring the amplitude, averaging, and calculating the square root of the result.

Sample: Instantaneous representation of an acoustic waveform by a discrete signal.

Sample rate: Number of samples captured by the ADC per second.

Sampling theorem: When a continuous signal of frequencies less than f is sampled at 2f, all of the information will be contained within the discrete signal. Also known as the Shannon sampling theorem/criterion or Nyquist sampling theorem/criterion.

Simultaneous masking: Masking which occurs when the maskee and masker components occur at the same time.

Signal to noise ratio (SNR): Ratio of the desired signal to the residual noise floor, expressed in dB.

Sound synthesis: Creation of sound through artificial means.

Speaker verification: Process of comparing the voice from a single speaker with another voice to verify that they are from the same person. Used in applications such as telephone banking.

Spectrogram: Visual representation of the magnitude of spectral information contained within a signal as a function of time.

Spectrum: Frequency components of a signal.

Speech recognition: Process of identifying words for applications, such as speech to text.

Splicing: Addition of a region of audio samples into a recording, either through cloning a region within the same recording or copying and inserting from a different recording.

Standard operating procedure (SOP): Documentation related to a laboratory process to allow an examiner of equal competence to perform the procedure, thus ensuring repeatability and reproducibility.

Stereo: Technique to create a spatial sound impression through the playback of audio over two or more channels.

Stereo encoding: Perceptual encoding of stereo signal information to reduce the amount of data required for the storage or transmission.

Stopband: Frequency components which are attenuated during a filter process.

Storage media: Physical object to store digital data.

Technical review: Evaluation conducted by a second qualified examiner of reports, notes, data, conclusions, and other documents.

Temporary Threshold Shift (TTS): Condition caused by over-exposure to sound which results in a temporary upwards shift of the hearing threshold and subsequently reduced sensitivity of the human auditory system.

Testimony: Declaration or statement given by a witness in legal proceedings.

Threshold of audibility (TOA): Minimum sound pressure of a pure tone which can be perceived by the human auditory system in the absence of other sounds.

Timbre: Overall tonal perception of a sound.

Time Domain: Signal in which time is the independent variable. Also known as the temporal domain.

Transcoding: Conversion between formats or encoding methods.

Transducer: Device used to convert one type of energy into another, such as a microphone or speaker.

Transient: An abrupt and short change to the spectral content of a signal.

Trimming: Process of removing samples positioned at the extremes of a signal.

μ: law (sometimes written mu:law): Companding technique to optimise the dynamic range of a system by reducing it from 14 bits to 8 bits of logarithmic PCM data. Used in North America and Japan for telephony communication.

Validation: Confirmation through objective evidence (often experiments), that the requirements for a specific intended use of a tool, technique, or procedure have been fulfilled.

Validation testing: Experiments to establish the efficacy and reliability of a tool, technique, or procedure.

Verification: Confirmation that a tool, technique, or procedure continues to performs as expected.

Voice-activated: Function found in a number of recording devices in which recording is activated once the magnitude of the sound passes a threshold.

Voice comparison: Process of comparing voices, commonly one of which the identity is unknown against a known set.

Voice Over Internet Protocol (VOIP): Technology that uses an internet connection for voice calls.

Waveform: Graphic representation of the magnitude of a signal as a function of time.

Waveform file format (WAVE): Standard audio format developed through a collaboration between IBM and Microsoft which commonly stores audio in an LPCM format (although audio of other formats can be stored).

Window function: Function applied to a frame to prevent artefacts at the edges when concatenated in post-processing.

Windows Media Audio (WMA): Collection of audio codecs created by Microsoft, supporting both lossy and lossless encoding.

Working copy: Bit-stream copy of a recording used for subsequent processing or analysis.

Workstation: Computer, including the associated hardware and software, used for forensic casework.

Write blocker: Hardware or software tool which prevents the modification of media content on external devices or storage media.

Zero crossing: Occurrence of a signal value intercepting the x-axis within a waveform representation of the signal.

Zero padding: Appending of zero value samples to a signal to meet criteria such as window lengths and to increase FFT resolution.

Index

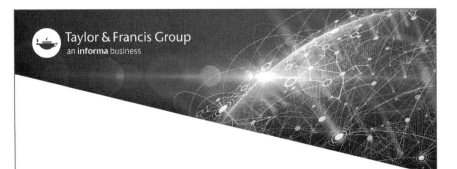